Praise for *Business Analytics with Management Science Models and Methods*

"As Business Analytics has become a popular topic in recent years, a number of texts on the subject have appeared in the market. However, most of these books simply present a collection of topics in data mining, statistics, and management science tools. Dr. Asllani's book has a refreshing new approach to business analytics—a logical flow of design thinking for decision support with management science methods. This book emphasizes the creative thinking approach to decision making through practical, intuitive, and real success application examples. This is an excellent text for students and practitioners of business analytics."

—**Sang M. Lee**, PhD, University Eminent Scholar Emeritus, University of Nebraska-Lincoln

"Dr. Asllani illustrates the relevance of management science in the era of Big Data and Business Analytics. He demonstrates how predictive analytics can inform and enhance prescriptive analysis, and how the rapid growth in computing power has impacted tackling larger optimizations. It is a great primer for someone new to the topic, and a great reference to anyone in the field.

After 10 years practicing management science and prior graduate level coursework, I have found that the content in Dr. Asllani's book has affected my professional modeling with a rigor and understanding that I didn't realize had been lacking.

The book is well-written and paced, and each chapter builds on concepts from the prior. End-of-chapter questions challenge the reader to recall information from the chapter and consider its practical applications."

—**Brett Senentz**, Business Optimization and Analytics Project Manager, McKee Foods Corporation

"Dr. Asllani has delivered a practical guide for practitioners in the field and a priceless textbook for students with one brilliant stroke. This book is certain to serve as an invaluable reference in analytics and management science. The book covers a wide array of applications, from production, to logistics, to marketing.

Dr. Asllani explains the intuition behind the concepts, avoiding heavy formulas and definitions, thus allowing for a guaranteed, solid grasp of each concept. He provides spreadsheet templates, which allow for easy application and reuse for a variety of optimization models. His step-by-step methodologies are sure to make the LP formulation process easier to apply by practitioners."

—**Alireza Lari**, PhD, Professor of Practice of Management, Wake Forest University School of Business

Business Analytics with Management Science Models and Methods

Arben Asllani

Professor of Business Analytics,
University of Tennessee at Chattanooga

Associate Publisher: Amy Neidlinger
Executive Editor Jeanne Glasser Levine
Operations Specialist: Jodi Kemper
Cover Designer: Alan Clements
Managing Editor: Kristy Hart
Senior Project Editor: Lori Lyons
Copy Editor: Karen Annett
Proofreader: Sarah Kearns
Indexer: Ken Johnson
Compositor: Nonie Ratcliff
Manufacturing Buyer: Dan Uhrig

Pearson Education, Inc.
Upper Saddle River, New Jersey 07458

For information about buying this title in bulk quantities, or for special sales opportunities (which may include electronic versions; custom cover designs; and content particular to your business, training goals, marketing focus, or branding interests), please contact our corporate sales department at corpsales@pearsoned.com or (800) 382-3419.

For government sales inquiries, please contact governmentsales@pearsoned.com.

For questions about sales outside the U.S., please contact international@pearsoned.com.

First Printing November 2014

ISBN-10: 0-13-376035-9
ISBN-13: 978-0-13-376035-4

Pearson Education LTD.
Pearson Education Australia PTY, Limited.
Pearson Education Singapore, Pte. Ltd.
Pearson Education Asia, Ltd.
Pearson Education Canada, Ltd.
Pearson Educación de Mexico, S.A. de C.V.
Pearson Education—Japan
Pearson Education Malaysia, Pte. Ltd.

Library of Congress Control Number: 2014948668

To my family, for their support and patience while I was writing this book.

Contents

Acknowledgments

I want to thank my mentor, Professor Sang M. Lee, for introducing me to the field of management science. His seminal work in goal programming inspired me to carry the torch of knowledge and love of learning. I also thank Professor Marc J. Schniederjans for teaching my first management science course and for guiding me throughout this book. His continued feedback and support is much appreciated. I also want to acknowledge my colleagues J. R. Clark, Lawrence Ettkin, Richard Becherer, and Michael Long for offering valuable and practical advice on how to approach the book. Finally, I want to recognize Jeanne Glasser Levine and the Production team at Pearson for working closely with me and providing continued support to make this book a reality.

About the Author

Arben Asllani is Marvin E. White Professor of Business Analytics at the University of Tennessee at Chattanooga. He has an M.A. and Ph. D. from the University of Nebraska at Lincoln and a B.S. degree from the University of Tirana, Albania. Dr. Asllani has been a member of the Decision Sciences Institute since 1997 and has joined several other traditional and online academic and practitioner oriented conferences and organizations. He has won several faculty teaching and research awards and is a member of Alpha Honor Society at the University of Tennessee at Chattanooga. Dr. Asllani is Associate Editor of the *American Journal of Business Research* and serves on the editorial board of *Service Business*. Dr. Asllani has published more than 36 articles in journals including *Omega, Transfusion, European Journal of Operational Research, Knowledge Management, Computers & Industrial Engineering, Total Quality Management and Business Excellence,* and *Service Business: An International Journal*. He has also published and presented over 30 research papers at academic conferences.

Dr. Asllani has a broad expertise in business analytics, especially in optimization techniques and computer-based simulations. He has served as a consultant and trainer to a variety of business and government agencies. Dr. Asllani has also taught extensively in management science, business analytics, and information systems courses, and has played an important role in developing business analytics programs in the United States and abroad.

Preface

Business managers have always used data to make decisions and gain a better understanding of business operations, products, services, and their customers. Today, however, the business world has entered the era of Big Data and the nature of organizational data has changed significantly. Big Data is characterized by high volume, variety, and velocity [1] and presents unique opportunities and challenges for the practitioners of management science.

Ninety percent of all data in the world has been generated over the last two years.[2] Every hour, Walmart handles more than one million customer transactions and generates over 2.5 petabytes of data. [3] This amount of data and information is continuously saved in digital storage, which has become increasingly less expensive as predicted by Moore's law. This famous law, which has influenced many aspects of information technology and electronics for almost five decades, can also be used to govern the era of Big Data. The law indicates that the amount of data available almost doubles every two years.[4]

The change from traditional to Big Data has caused significant changes in the field of data analytics. Thomas Davenport [5] describes three types of analytics: Analytics 1.0, Analytics 2.0, and Analytics 3.0. According to Davenport, the business intelligence, also known as Analytics 1.0, was replaced in the mid-2000s with Analytics 2.0. Today, business organizations have entered the era of Analytics 3.0, which "powers consumer products and services" with data and information [5, p. 65]. One of the requirements for capitalizing on Analytics 3.0 is prescriptive analytics. Davenport writes:

> There have always been three types of analytics: descriptive, which reports on the past; predictive, which uses models based on the past data to predict the future; and prescriptive, which uses models to specify optimal behaviors and actions. Although Analytics 3.0 includes all three types, it emphasizes the last. Prescriptive models involve large-scale testing and optimization and are a means of embedding analytics into key processes and employee behaviors [5, p. 70].

About the Book

This book is about prescriptive analytics. It provides business practitioners and students with a selected set of management science and optimization techniques and discusses the fundamental concepts, methods, and models needed to understand and implement these techniques in the era of Big Data. A large number of management science models exist in the body of literature today. These models include optimization techniques or heuristics, static or dynamic programming, and deterministic or stochastic modeling. The topics selected in this book, mathematical programming and simulation modeling, are believed to be among the most popular management science tools, as they can be used to solve a majority of business optimization problems. Over the years, these techniques have become the weapon of choice for decision makers and practitioners when dealing with complex business systems.

Business systems are typically complex, and as a result, decision makers must incorporate many variables when utilizing optimization or simulation models. The process of creating and solving such models is further complicated when large amounts of data must be incorporated, as in the case of Analytics 3.0. The book offers several models and methods that emphasize the practical aspects of management science in the era of Big Data. Templates, algorithms, and user-friendly interfaces are included to make Big Data decision models more practical and easier to implement.

The book is organized in ten chapters. Chapter 1, "Business Analytics with Management Science," discusses the role of management science in the era of Big Data. The importance and scope of business analytics in today's organizations are described briefly. The chapter concludes with a discussion of several challenges faced by today's organizations when implementing business analytics and management science models.

Chapter 2, "Introduction to Linear Programming," demonstrates the importance of linear programming (LP) as a business analytics tool and discusses the potential use of LP to improve organizational performance. The models are explained graphically, and their solutions are demonstrated graphically and via the Microsoft Excel Solver add-in. The focus is on the understanding of fundamentals of mathematical programming in general and linear programming in particular. The graphical approach used in this chapter serves as the basis for intuitive explanations of more complex models covered in the remaining chapters.

Chapter 3, "Business Analytics with Linear Programming," explores more advanced models of LP, models that include many decision variables and constraints. Special focus is placed on the data-input requirements and input variables of linear programming models. Practical recommendations related to linear programming models and its use for data analytics are discussed in this chapter. The chapter also provides a detailed discussion of Answer Reports and sensitivity analysis. The concept of reduced costs, shadow prices, and upper and lower limits are discussed in detail and their role in decision insights is highlighted.

Chapter 4, "Business Analytics with Nonlinear Programming," discusses the use of nonlinear programming (NLP) models for business analytics. The difference between LP and NLP models is highlighted, and several areas of application of the NLP models in real business settings are recommended. The chapter also discusses challenges of using NLP models and offers practical recommendations for the management scientists. Formulation methodology and solution steps are demonstrated via examples and their respective Excel templates.

Chapter 5, "Business Analytics with Goal Programming," involves goal programming (GP), a powerful programming tool with multiple objectives. Utilizing the knowledge from LP models, the chapter offers an intuitive explanation of GP modeling and discusses different approaches to solve GP. Simple multigoal decision-making examples are used to demonstrate the essence of goal programming and Solver is used to demonstrate their solutions.

Chapter 6, "Business Analytics with Integer Programming," briefly covers another type of mathematical programming, integer programming. Integer LP models are similar to LP models with the additional constraints of having the integer values for decision variables. Integer GP models are similar to GP models with the additional constraints of having the integer values for decision variables. The use of Solver makes the solution methodology of integer programming very similar to the ones explored in the previous chapters (LP and GP). In addition, Chapter 6 discusses two special types of integer programming models: the Assignment Method and the Knapsack Problem.

Chapter 7, "Business Analytics with Shipment Models," discusses the use of mathematical programming models in shipment and logistics. Decision makers today seek to optimize not only production or service operations, but also those operations that transport goods from plants, to warehouses, to distribution centers, or to other destinations. The transporting of such goods includes organizational, intraorganizational, and interorganizational shipments and accounts for a significant part of the costs of products and services. Chapter 7 discusses two types of shipment problems: transportation and transshipment models.

There are complex LP models with many constraints and decision variables that use relatively small amounts of data. But there are also simple LP models with only a few constraints and decision variables that use large input (Big Data) sets. Chapter 8, "Marketing Analytics with Linear Programming," demonstrates the application of simple LP models using information generated from customer relationship management systems (CRMs). CRMs are organizationwide systems that, among other things, can be used to store sales information, such as customers, transaction dates, and amount of sales in each transaction. Using a case-based approach, this chapter utilizes the recency-frequency-monetary value (RFM) analysis and combines it with an LP model. This combination of Big Data with an appropriate LP model can be used to generate recommendations for effective marketing campaigns.

Chapter 9, "Marketing Analytics with Multiple Goals," also uses a case-based approach to apply more advanced programming models in the RFM analysis. These models combine several dimensions of the RFM analysis and model marketing campaigns with multiple objectives via linear and goal programming.

Chapter 10, "Business Analytics with Simulation," demonstrates the use of simulation for data analysis. This section provides an intuitive explanation of computer simulation. Examples are provided and model assumptions are discussed. Several areas of application of the model in real business settings are also recommended. The chapter offers a step-by-step approach that practitioners can use to design and build simulation models in business settings. The methodology is illustrated with an example and Excel as an analysis tool.

The book concludes with two appendixes. Excel spreadsheet templates are used to support modeling approaches offered in the book. As such, a variety of Excel tools, which are used throughout the book and must be included in every data scientist's toolbox, are offered in Appendix A, "Excel Tools for the Management Scientist." It is advised that the reader consult the material in Appendix A before or during the exploration of models presented throughout the chapters. Finally, Appendix B, "A Brief Tour of Solver," is a tutorial on Solver, the Excel add-in used in this book. Although Solver is sufficiently explained in each chapter, the reader can use Appendix B for further explorations of the features of this add-in.

Features of the Book

This book provides a unique approach to management science. The following features facilitate a better understanding of the discussed models and allow for a practical implementation of these models in real-world business situations.

- **Prescriptive Analytics in action**—Each chapter starts with a description of an application of a management science topic in a real business environment.

- **Management Science and Big Data**—The book describes management science models in the context of business analytics. Special focus is placed on data processing and other challenges presented by the high volume, velocity, and variety of model inputs. Each chapter ends with a discussion of challenges posed by Big Data in the management science topic and its practical implications.

- **"Black box" approach**—The topics discussed in this book focus more upon the input-output aspects of decision making and less upon the dynamics and complexities of the model itself, which is handled by Solver.

- **Explaining by example and intuition**—Complex concepts are often understood via examples and intuitive explanations, not by formulas and theoretical definitions. Several examples are used throughout the book. The graphical approach discussed in Chapter 2 builds the foundation of explaining difficult concepts of management science with intuition and by example.

- **Practice problems**—Each chapter concludes with a series of conceptual questions that further add to the understanding of the topics. In addition, several "end of chapter" problems provide a thorough training opportunity to the reader. Problems require not only problem formulations and solutions, but also necessitate that the reader deal with Big Data sets and real-life challenges. Additional problems are provided on the book's companion website.

- **Excel templates and data sets**—A series of templates used in the book and several data sets for additional problems are available to students, practitioners, and instructors on the book's companion website at www.informit.com/title/9780133760354.

- **Instructor's resources**—The book's companion website also contains all necessary resources for the instructors, such as the complete instructor's solution manual, PowerPoint lecture presentations, and the test item file.

1

Business Analytics with Management Science

Chapter Objectives

- Emphasize the importance of business analytics in today's organizations
- Discuss the scope of business analytics and the set of skills required for business analyst practitioners
- Discuss the role of management science models in the Big Data era
- Explain challenges faced by organizations when implementing business analytics
- Examine the new role of management science in the era of Big Data

Prescriptive Analytics in Action: Success Stories

There is no doubt that organizations have become more competitive through the use of business analytics. A 2013 study by the *MIT Sloan Management Review* indicates that 67% of companies use data analytics to gain a competitive advantage compared with only 37% in 2010.[6] Almost any business—be it focused on consumer products, the entertainment industry, healthcare, or fast food—is using mountains of data to improve customer service, operations, supply chains, or product designs.[7] Business analytics is no longer a buzzword. A recent survey of Gartner Inc. conducted in June 2013 indicates that over 30% of companies in each sector, such as media and communications, banking, and services, have already invested in data analytics.[8] The same survey stated that over 50% of companies in transportation plan to invest in the next two years in data analytics. The number of investors in data analytics is currently 41% in health care, 40% in insurance, 39% in retail, and 38% in government.[8]

The analytics can be implemented in almost all business functions. Currently, according to the survey, almost 55% of respondents indicated that business analytics is used to improve customer experience, 49% use business analytics to increase efficiency, 42% use it for marketing purposes, and 37% use it for cost reduction.[8] In the same survey, 70% of respondents indicated the use of business transactions as the source of data; 55% use log data, 42% use machine sensors, 36% use source e-mails and documents, and 32% utilize social media.

The use of business analytics for productivity improvement has expanded not only to large corporations but also to smaller companies. First Tennessee Bank, for example, lowered its marketing cost by 20% and increased its return on investment by more than 600% by using data analytics for better customer marketing.[9] The following bulleted items briefly discuss how Target and LinkedIn use business analytics to reach more customers:

- *Target—Retailers have collected customer information for decades. This information has helped retailers to increase their sales and create targeted promotions based on specific customer segmentation. However, Target has moved data analytics into a new stage. Instead of advertising to customer segments, Target now is able to "laser target" each customer with his or her specific needs for specific products. In early 2012, the* New York Times *reported the story of a Target analyst who was able to determine if a customer was pregnant based on the pattern of previous purchases. That information was then used to advertise targeted products.[10] Target's sales skyrocketed, and the company's revenues have grown by $23 billion since the new data-informed strategy was implemented.[11]*

- ***LinkedIn***—*LinkedIn, the social networking website, was founded in December 2002 and was first launched in May 2003. In 2006, the data scientists at LinkedIn started to investigate connection patterns and profile richness of their 8 million users. They tested what would happen if a user was presented with names of people with whom they had not yet connected but seemed likely to know. And, it worked. "People you may know" ads achieved a 30% higher click-through rate than the rate obtained by other prompts [12] and the campaign generated millions of new page views and the LinkedIn membership network grew significantly. In May 2014, LinkedIn had 300 million registered users worldwide.[13]*

Introduction

The automation of business processes with information technology has led to the automatic capture of massive data. Successful managers must be able to work with this data and make sense of all this information. They must understand not only the math and algorithms, but also the "big picture approach to using Big Data to gain insights."[7] Business analytics has become their weapon of choice.

What is business analytics? There are many definitions of business analytics as a field of study. From a practitioner's perspective, business analytics can be defined as set of tools and techniques that are used to retrieve, process, transform, and analyze data in order to generate insights for better decisions in the business world. Wayne Winston, a prominent scholar and consultant in management science and prescriptive analytics, defines analytics as simply "using data for better decision making."[14]

Business analytics integrates tools and techniques from four major fields: information management, descriptive analytics, predictive analytics, and prescriptive analytics. Information management deals with storing, extracting, transforming, and loading data and information from operational databases into data warehouses. Once the information is made available in data warehouses and data marts, business analysts can use a series of descriptive analytics tools to understand what has happened in the organization regarding its key performance indicators.

Further, predictive analytics tools can be used to forecast and estimate future behavior based on past performance. Finally, optimization and other management science models are used as prescriptive analytics to identify the best courses of actions and optimal decisions. The nature of management science has changed to accommodate the need to process large amounts of data sets. An obvious feature of today's management science is its heavy reliance on spreadsheet modeling and other data analysis software programs.

Implementing Business Analytics

As organizations are becoming more competitive by using business analytics, there is no doubt that practitioners, from both large and small companies, are eager to learn more about data analytics and how to implement it in their everyday decision making. But how do managers implement business analytics in the workplace? Is there a methodology or a recommended set of steps that can be followed by practitioners? The examples of Target, LinkedIn, First Tennessee, and many other companies that have implemented data analytics can be used to derive practical steps for implementing analytics in organizational settings. Business analytics initiatives in the era of Big Data usually follow an eight-step cycle:

1. Understand the company's products in depth.
2. Establish tracking mechanisms to retrieve the data about the products.
3. Deploy good-quality data throughout the enterprise.
4. Apply real-time analysis to the data.
5. Use business intelligence to standardize reporting.
6. Use more advanced analytics functions to discover important patterns.
7. Obtain insights to extract relevant knowledge from the patterns.
8. Make decisions to derive value using the knowledge discovered.

These eight steps illustrate how organizations utilize all aspects of data analytics. LinkedIn, for example, uses information about its members, which is housed in operational databases. This information is then organized into a data warehouse and the analysts can use this information to explore the browsing history of LinkedIn members. Furthermore, LinkedIn uses descriptive statistics to generate reports and discover patterns. These patterns let the data analytical team at LinkedIn use predictive analytics to discover that speed is very important in receiving positive responses. Specifically, LinkedIn analysts were able to determine that "adaptation exponentially increases as the response time goes towards sub-seconds." Finally, prescriptive analytics is used to generate appropriate actions. For example, LinkedIn could use optimization techniques to identify the best mix of companies or individuals, which maximizes the number of prospects or product sales.

Business Analytics Domain

As shown in Figure 1-1, the domain of business analytics covers four major areas of study: databases and data warehouses, descriptive analytics, predictive analytics, and prescriptive analytics. Whereas data structures are used to effectively store and efficiently retrieve information, descriptive analytics can be used to report the past. Whereas predictive analytics uses past data to create models that predict the future, prescriptive analytics utilizes optimization, heuristics, or simulation models that can specify optimal solutions and prescribe the best courses of action.

Databases and Data Warehouses

Databases and data warehouses serve as the foundation of business analytics. Every business analytics process starts by storing the data appropriately in operational databases and ensuring data integrity. The data analyst must understand the principles of database design and implementation throughout all its steps: conceptual, logical, and physical modeling. The most common design of databases is known as relational database modeling. Relational databases are distributed throughout organizations and may belong to different departments. They may be stored in different platforms and may use incompatible data formats.

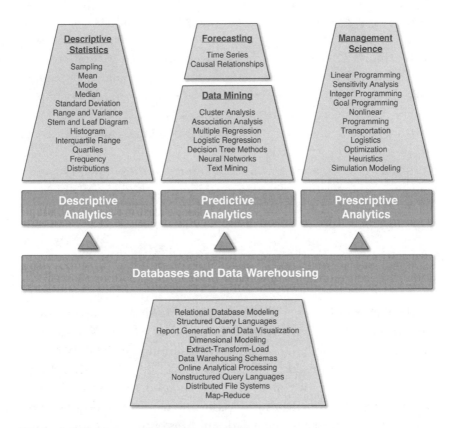

Figure 1-1 Overview of business analytics

Data warehouses consolidate information gathered from disparate sources and provide access of customized information to business users so they can make better decisions. It is important that data stored in disparate sources is loaded into the data warehouse in a consistent format. In addition, data warehouses combine structured, semi structured, and nonstructured data. In addition, extract-transform-load (ETL) processes can also be used to refresh and update operational databases into the warehouse target objects. Data warehouses are then available for queries through multidimensional objects such as cubes.

After the information is captured and stored in operational databases or data warehouses, the data analyst may perform online analytical processing, create business reports, visualize data, and produce operational business intelligence. Structured Query Language (SQL) is a programming language that can be used to create databases, store and update data in these databases, and retrieve information from them.

The nature of data capturing and processing has changed dramatically in the era of Big Data. Nonrelational, distributed, open-source, and horizontally scalable (abbreviated as NoSQL) databases have emerged and are used in real-time web applications.[15] Whereas relational databases consist of related tables, records, and fields, NoSQL databases contain nonstructured data in the form of key-values, graphs, or documents.

Descriptive Analytics

Descriptive analytics is used to quantitatively describe the main features of organizational data. Descriptive analytics aims to summarize a sample, rather than use the data to learn about the population that the sample of data is thought to represent. Some of the common tools used in descriptive statistics include sampling, mean, mode, median, standard deviation, range, variance, stem and leaf diagram, histogram, interquartile range, quartiles, and frequency distributions. The results of descriptive statistics are often displayed via graphics/charts, tables, and summary statistics such as single numbers.

Suppose that Fandango, the leading online ticket seller for movie theaters, wants to investigate the movie preferences of its customers during the past year. Fandango sells millions of tickets to approximately 20,000 movie theaters across the United States.[16] Information about customers, movie theaters, ticket sales, and show times are automatically captured and stored in structured databases. Then, periodically, this information is extracted, transformed, and loaded into data warehouses or data marts, which mostly reside in distributed servers. Fandango data scientists will then use descriptive analytics. For example, using a sample of movie titles, the analysts can investigate the correlations among total sales for different movies. Using a sample of moviegoers, they can calculate the average ticket sales for a week, the most popular movie, distribution of customers among movie genres, the busiest hours of the day in the movie theater, age distribution of moviegoers, gender distribution, and so on. This type of data analysis helps Fandango set ticket prices, offer discounts for certain movies or show times, and assign show times of the same movie in different theaters.

Predictive Analytics

Whereas descriptive statistics are considered a straightforward presentation of facts, predictive analytics uses statistical modeling to draw conclusions and predict future behavior based on the assumption that what has happened in the past will continue to happen in the future. Some of the common tools used in descriptive statistics include cluster analysis, association analysis, multiple regression, logistic regression, decision tree methods, neural networks, and text mining. Forecasting tools, such as time series and causal relationships, are also classified as predictive analytics.

How does Fandango know to send e-mails to its members with discount offers for a specific movie on a specific day? Predictive analytics tools can crunch terabytes and terabytes of data to determine that while John likes science fiction movies, he has not seen the latest sci-fi movie, which has been in the theaters since last Friday. How does a grocery store checkout system generate valuable coupons just in time and on the back of the printed receipt? Julie's favorite whole-grain cereal was missing from the shopping basket that day. The computer matches Julie's past cereal history to ongoing promotions in the store, and right there, on the spot, Julie receives a coupon for the whole-grain cereal that she will most likely buy.

Prescriptive Analytics

Some of the most common models used in prescriptive statistics include linear programming, sensitivity analysis, integer programming, goal programming, nonlinear programming, and simulation modeling. Practitioners use prescriptive analytics to make decisions based on data. For example, continuing with the Fandango example, the prescriptive tools allow for ticket price offerings to change every hour. Fandango has learned when the most desirable movie times are by sifting through millions and millions of show times instantaneously. This information is then used to set an optimal price at any given time, based on the supply of show times and the demand for movie tickets, thus maximizing profits.

Prescriptive analytics can help the movie industry to ensure that their pricing structures are optimally set to contribute to bottom-line results. Similarly, prescriptive analytics can help airline industries maximize their revenues by making sure that the highest prices are charged during the highest times of demand as well as by lowering the prices when the demand is low. The combination of Big Data with prescriptive tools allows the airlines to adopt pricing policies that go beyond traditional peak, off-peak, or shoulder seasons. Changes are dynamic and in real time; they can be implemented within the days of the week or even the hours of the day. Prescriptive analytics are the engine of today's real-time business intelligence.

Challenges with Business Analytics

Optimization modeling and heuristic tools have yet to make the transition into the Big Data era. In an October 2013 issue of the *Wall Street Journal*, John Jordan of Penn State University describes several challenges involved when implementing business analytics. [17] He notes that there is "a greater potential for privacy invasion, greater financial exposure in fast-moving markets, greater potential for mistaking noise for true insight, and a greater risk of spending lots of money and time chasing poorly defined problems or opportunities." This section discusses some of the challenges associated with prescriptive analytics and offers some practical recommendations on how to avoid them.

Lack of Management Science Experts

The everyday use of mathematical modeling and other techniques requires that business managers or other practitioners have a good understanding of numeracy and mathematical skills. However, there is a lack of such skills, especially for medium-sized or small organizations. It is estimated that by 2018, U.S. universities and other educational institutions will need to produce between 140,000 and 190,000 more graduates for deep analytical talent positions and 1.5 million more data-savvy managers.[18]

Business analytics, in general, and prescriptive analytics, in particular, can become more "popular" with the use of spreadsheet modeling. Spreadsheet modeling is widely used in colleges and universities for teaching mathematical programming. Instead of heavy modeling, which seeks optimal solutions, spreadsheet modeling techniques include simpler formulations, which seek practical solutions. However, the spreadsheets have limitations in the amount of data they can store. They cannot store data about millions of transactions in a bank or the details of federal spending on transportation projects, even for a week. There is a time for organizations to introduce more advanced tools.

Analytics Brings Change in the Decision-Making Process

The goal of prescriptive analytics is to bring business value through better strategic and operational decisions. At the strategic level, those who make decisions about what models to implement and what needs to be measured will accrue more power. At the operational level, the implementation of such models brings a power shift in the decision-making process. Information-based decisions across organizational boundaries can upset traditional power relationships.

The story of the Oberweis Dairy [19] is an excellent example of how data analytics can transform organizations. The company started as an Illinois farmer selling his surplus milk to neighbors in 1915. Today, the company has "three distribution channels: home delivery, with thousands of customers; retail, with 47 corporate and franchise stores; and wholesale, to regional and national grocery chains like Target." The usual approach to decision making at the company is by asking executives to figure out the best configuration for future changes.

In 2012, the company was trying to expand its operations geographically, and the chief executive officer (CEO) asked a data analytics executive with only three years of experience to join the strategy table. Bringing analytics to the table changed the preconceived notion about the customers. Although there were current customers who benefited from the offerings of the company, data analysis indicated that the company had spent a lot of money to acquire customers who should not have been approached in the first place. Contrary to the company's conventional wisdom, data from the customer sales indicated that "the so-called Beamer and Birkenstock group—liberal, high-income, BMW-driving, established couples living leisurely lifestyles" was not a good fit for the dairy farm. So, the meeting changed from a tactical meeting with a focus on "how many trucks and transfer centers would be required" into a strategic "define the target market" meeting. The change is cultural, and it has grown to a point now where people want to acquire a better understanding of analytics tools because they can see that there is real benefit.

Big Data Leads to Incorrect Information

Modeling with business analytics is more an art than a science. One fundamental step when building mathematical models is the process of abstraction. Through this process, the modeler eliminates or suppresses any unnecessary details and allows only the relevant information to enter the model. When good information goes in the model, a good model produces good results. The opposite is known as GIGO (garbage in, garbage out). In the era of Big Data, it is significantly more difficult for the data analyst to mine in the mountains of information and find the relevant pieces.

Very often, valid models produce poor results, which lead to the wrong decisions. In the era of Big Data, this happens very often. A recent story [20] reports how ten volunteers checked the accuracy of their information on AboutTheData.com and they each found inaccuracies. In one specific case, a volunteer found that "she had two teens, at 26." Interestingly, a CNN team found that Acxiom, the company that runs the database, was more accurate specifying the interests and less accurate in demographic data (marriage status, number of children). Wrong assumptions can lead to wrong decisions. If you are a company purchasing this database, you know the interests of your future customers, but very likely you may be sending out 2 million direct mails pieces on baby products to people who may not even have children.

Big Data Demands Big Thinking

Business organizations are just entering the new paradigm of Big Data. They have been using standard databases for more than three decades and have accumulated experience and knowledge. However, Big Data demands new techniques and many of them are still in the developmental stages. Acquiring the new tools requires a radical change in underlying beliefs or theory—they require a new way of thinking. It requires, for example, that more people think probabilistically rather than anecdotally. It also requires that managers learn to focus on the signals and do not get lost in the noise.[21] This way, organizations will be able to better understand the factors behind customers, products, services, and how to make analytical decisions.

What is Big Data? Big Data is a combination of structured, in-house operational databases with external databases, with automatically captured and often nonstructured data from social networks, web server logs, banking transactions, content of web pages, financial market data, and so on. All this data, coming from a wide variety of sources is combined into non-normalized data warehouse schema. Big Data is usually characterized by three Vs: volume, velocity, and variety:[1] [22]

- **Volume**—Today, the high volume of business transactions is automatically captured by advanced enterprise information systems. Nonstructured and external databases also produce large amounts of data. These sources are then combined into denormalized data warehouses. Unnormalized (or denormalized) data means high-volume data with intentional redundancy. The volume of Big Data is larger than the volume processed by conventional relational databases.

  Descriptive and predictive analytics benefit from the high volume of data. After all, statistical analysis and reliability of predictions is better when the population size increases. A forecasting method with hundreds of factors can predict better than the one with only a few input factors. Prescriptive models also benefit from Big Data. They are based on aggregated inputs, which are the result of descriptive analytics: contribution coefficients, average processing times, mean of distributions, and so on. The validity of these aggregate values improves with high-volume data.

Stochastic models can benefit from high-volume data, as well. Statistical distributions are more reliable when fitted with a large number of data points. A prescriptive model, which assumes a normal distribution for processing times, is more reliable when the mean and standard deviation of the normal distribution is based on thousands or millions of data points as compared with only hundreds of data points.

- **Velocity**—Velocity refers to the rate at which data flows into an organization. Online sales, mobile computing, smartphones, and social networks have significantly increased the information flow for the organization. Organizations can analyze customer behavior, sales history, and buying patterns. They are able to quickly produce operational business intelligence and recommend additional purchases or customized marketing strategies. The velocity of system output is also important. The recommendations must be delivered in a timely manner and must be included as part of business operations. A loan officer, for example, could compare the information in a loan application against business rules and mining models, and make a recommendation to the applicant or make a decision about the loan.

  Prescriptive modeling techniques can take advantage of velocity. They can be modeled to run in the background and take data from input to make optimal or near-optimal decisions.

- **Variety**—Variety of data refers to the mix of different data sources in different formats. As mentioned earlier, Big Data input may arrive in the form of a text from social networks or an image from a camera sensor. Even when the data source is structured, the format can be different. Different browsers generate different data. Different users may withhold information, or different vendors may send different information based on the type of software they use. Of course, every time humans are involved, there may be errors, redundancy, and inconsistency. Management science models require the input data to be uniform. As such, the implementation of these models in the era of Big Data normally requires an additional layer between the source data and the prescriptive model.

Exploring Big Data with Prescriptive Analytics

Big Data brings both opportunities and challenges for prescriptive analytics. The *volume* generally improves the quality and accuracy of management science models. Good management science models can speed up the flow of information by offering quicker decisions and improving operational business intelligence. *Variety*, though, is seen as a hindrance to the implementation of management science techniques. However, with the right technological framework, the negative impact of *variety* can be mitigated.

Information technology (IT) is probably the most important ingredient in the Big Data recipe. Over the last decade, changes in IT have brought significant changes in the nature of Big Data. In the past, companies would build database applications and plan server capacity just enough to store a few years' worth of data. Beyond that, companies would simply delete the transaction data that was older than three to five years. Today, data storage costs are relatively inexpensive, so companies can afford to store data and information generated by these transactions.

Automatic identification and data capture technologies are rapidly developing, and they allow for faster identification of objects, better collection of data about them, and automatic entering of that data directly into computer systems without human involvement. Such technologies include bar codes, radio frequency identification, magnetic stripes, biometrics, optical character recognition, smart cards, cameras, and voice recognition. Finally, laws and regulations are approved for data storage, processing, security, transparency, and auditing (for example, the Sarbanes–Oxley Act of 2002 [23]), which make it mandatory for organizations to store the information. All these factors have contributed to the rise of the era of Big Data.

Table 1-1 Challenges of Big Data to Optimization Models

Big Data Dimension	Challenges	Technology-Based Solutions	Methodology-Based Solutions
Volume	Managing large and rapidly increasing data sources	Advanced software programs able to process large number of constraints and decision variables	Standardize the ETL processes to automatically capture and process input parameters
			Encourage system-driven versus user-driven optimization programs
Variety	Dealing with heterogeneity of data sources Dealing with incomplete data sets	Relational database systems and declarative query language to retrieve data input for optimization models ETL toward specialized optimization-driven data marts	Add data structuring prior to analysis Implement data cleaning and imputation techniques
Velocity	Managing large and rapidly changing data sets Reaching on-time optimal solutions for operational business intelligence	Advanced optimization software with the capability to reach optimal solutions within a feasible amount of time Optimization packages that directly connect to operational databases	Consider a trade-off between less-than-optimal but time feasible and practical solution and optimal but complex and often delayed solutions

Table 1-1 summarizes challenges of implementing optimization models in the era of Big Data and suggests conceptual approaches for management scientists to deal with them. High *volume* of Big Data requires that decision scientists have the capability to store and process a large amount of data. Cloud computing technology, which has risen over the past few years, has dramatically increased the ability for the businesses to store and process information. This technology offers dynamic and large distributed platforms for organizations to process input parameters and solve models at a large scale. These platforms can be used to run advanced optimization models, which engage multiple clusters. For example, advanced linear programming (LP) models have recently entered the paradigm of declarative programming. The goal of declarative programming is to ease the programmer's task by separating the control from the logic of a computation.[24] According to this paradigm, the set of LP constraints, for example, is solved by giving a value to each variable so that the solution is consistent with the maximum number of constraints. This allows for engaging multiple clusters when solving large optimization models and as such makes the implementation of large-scale and heavy computational models practically feasible to solve. The use of a declarative programming approach to model and solve mathematical programming models is still at an early stage.

Apache Hadoop is another good example of using advanced technology to handle high variety of data in optimization models. Hadoop, an open source platform, offers distributed computing, which places no condition on the structure of the data it can process. As such, Hadoop can be used as a great platform to mitigate the variety of components of Big Data. Google has introduced MapReduce,[25] a programming model approach that can process and filter (map), as well as merge and organize (reduce) large data sets according to specified criteria. MapReduce programs can be automatically and simultaneously executed across several computers, thus saving processing time while handling large amounts of input data.

Wrap Up

Management science techniques discussed in this book have been around for decades. However, in the era of Big Data, management scientists have "rediscovered their roots" [26] and are modifying traditional techniques to better process large volumes of data, offer simpler and practical models, utilize spreadsheet modeling techniques, and offer practical solutions, which can be implemented in real time. With the availability of large amounts of data, Big Data is the major factor that boosted management science into today's stage of prescriptive analytics.

Presently, several optimization software programs exist, which are able to model and to solve a large number of constraints and decision variables. Solver is an excellent program, licensed to Excel as an add-in from Frontline Systems. Solver can be used by practitioners to solve mathematical programming models and perform what-if analysis and optimizations to determine the best product mix, determine optimal shipping routes, maximize profit, or minimize costs. This book advocates a two-step approach when using Excel's Solver: (a) setting up a template and (b) running Solver and analyzing the results. This approach allows the modeler to design templates that handle large amounts of input data and then reuse these templates with new data sets as these sets are continuously updated from transaction database sources.

In the era of high-volume data, ETL processes can be used to automatically capture and process input parameters. These processes query transactional records and calculate averages for technological coefficients, contribution coefficients, and right-hand side values for up-to-date available resources. Automatic capturing and processing of data allows organizations to design optimization models that are process driven. Such an approach requires analytics to be embedded within business processes [27] and continuously adjust input parameters and periodically produce optimal solutions.

Considering these challenges, the content of the book is offered with two universal principles in mind: teaching by example and explaining by intuition. Practitioners understand complex concepts by referring to examples and understand the reasoning behind these concepts by consulting intuitive explanations, not by referring to formulas and theoretical definitions. The book uses a *black-box* approach, which allows the practitioner to focus more upon the input-output aspects of decision making and less upon the dynamics and complexities of the model itself, which in most cases are handled by software programs.

Review Questions

1. What is Big Data? Discuss the factors that led to the era of Big Data. Compare and contrast the paradigm shift from traditional operational databases to the Big Data analytical data warehousing.

2. What is business analytics? What is the difference between analytics and statistics? Briefly describe the domain of the four major fields of business analytics: databases and data warehousing, descriptive, predictive, and prescriptive analytics.

3. Compare and contrast descriptive, predictive, and prescriptive analytics in terms of tools and techniques used, data input and output, and their use in the decision-making process. Which type of analytics is more important for an organization?

4. The traditional management science techniques have been revitalized in the era of Big Data and have become the basis for many prescriptive analytics models. Explain the change in the nature of management science to accommodate the need to process large amounts of data.

5. The majority of companies today are using data analytics to gain a competitive advantage. Describe various ways in which business analytics can be used to lower costs, improve customer experience, and increase productivity. Make sure to support your argument with examples reported in newspapers, magazines, and online sources.

6. Select a company where you are a regular customer. Think of the potential data this company stores about you. Also, consider other data sources your company might use (for example, demographics, market information, consumer information such as credit score). How do you think the business uses this information to make you a more valuable customer? For example, how does your bank know to send e-mails to you with discount offers for a loan product? How does your favorite restaurant generate a free appetizer coupon to be used during your next visit?

7. Using the same company from the previous question, brainstorm ideas about how the company can improve its operations by lowering costs or improving productivity. For example, how does the bank identify the optimal interest rates for their new loan offerings? How does the restaurant decide the menu prices and happy hour discounts?

8. What is *volume* in the Big Data definition? How does the high volume of Big Data impact descriptive, predictive, and prescriptive analytics? Provide examples to illustrate your ideas.

9. What is *velocity* in the Big Data definition? How does high velocity of Big Data impact descriptive, predictive, and prescriptive analytics? Provide examples to illustrate your ideas.

10. What is *variety* in the Big Data definition? How does a high variety of Big Data impact descriptive, predictive, and prescriptive analytics? Provide examples to illustrate your ideas.

11. Discuss challenges that organizations face when trying to analyze Big Data. Make sure to include in your discussion one or more challenges such as privacy invasion, financial exposure, mistaking noise for the signal, and poorly defining business problems.

12. How important is it to have *optimal* versus *good but practical* solutions? Discuss the importance of spreadsheet modeling in this comparison. List other modeling software and compare it with spreadsheets.

13. Explore advantages and disadvantages of Excel modeling for prescriptive analytics. Consider in your discussion the ability of spreadsheets to process large amounts of data as well as the potential use of add-ins.

14. Good analytics models can sometimes lead to bad business results, conclusions, and recommendations. List at least three reasons why this might happen. For each reason, offer practical recommendations to avoid erroneous conclusions. Provide examples to illustrate your ideas.

15. Discuss challenges faced by practitioners when exploring Big Data with management science models. Suggest practical solutions to these challenges.

Practice Problems

1. A good analytics project starts with asking the right business questions. Consider a business organization, probably the one for which you are working, and state business questions whose

answers can help improve organizational performance. Then, focusing on a specific business function (such as marketing, operations, finance, accounting, human resources, etc.), explore the sources of data that can be used to answer these questions and discuss the potential use of such data for decision making.

2. Select an organization (such as Walmart, Amazon, your favorite restaurant, or even the one you work for) and discuss what type of information this organization might store and how data scientists can use that information for descriptive, predictive, and prescriptive purposes.

3. Create a table with three columns respectively named *Descriptive*, *Predictive*, and *Prescriptive,* and place each skill from the following list in the correct column:

Association analysis

Causal relationships

Cluster analysis

Decision tree methods

Frequency distributions

Goal programming

Histogram

Integer programming

Interquartile range

Linear programming

Logistic regression

Mean

Median

Mode

Multiple regressions

Neural networks

Nonlinear programming

Optimization heuristics

Quartiles

Range

Sampling

Sensitivity analysis

Simulation modeling

Standard deviation

Stem and leaf diagram

Text mining

Time series

Variance

4. The eight-step cycle to implement a business analytics program discussed in this chapter can serve as a template for any organization that is considering implementing a data analytics project. Using a specific organization (for example, from your own company), map a business analytics initiative into these eight steps.

Supposing you are a business analyst, offer potential specifics and details of each step, making reasonable assumptions.

5. Business analytics is being implemented by organizations in one or more business functions. Conduct an Internet search to find examples of data analytics projects implemented in specific functions. Place your findings in the respective cells of the following table.

Business Function	Examples of Descriptive Analytics	Examples of Predictive Analytics	Examples of Prescriptive Analytics
Operations			
Purchasing			
Marketing			
Finance			
Administration			
Human Resources			
Public Relations			
Add more functions here based on your findings...			

2

Introduction to Linear Programming

Chapter Objectives

- Discuss the importance of linear programming as a business analytics tool in business organizations
- Explore the components of linear programming models and relate them to business goals and constraints
- Provide a step-by-step formulation methodology for linear programming models
- Solve linear programming models graphically
- Solve linear programming models with the Excel Solver add-in
- Understand the constructs of linear programming modeling, its formulation, and solutions
- Discuss challenges of using linear programming models in everyday decision making

Prescriptive Analytics in Action: Chevron Optimizes Processing of Crude Oil[1]

Chevron is one of the world's leading integrated energy companies. Chevron refineries process crude oil into fuels, such as gasoline, jet, and diesel; base oils, such as lubricants; and specialty products,

---

[1] Source: https://www.informs.org/Sites/Getting-Started-With-Analytics/Analytics-Success-Stories/Case-Studies/Chevron

such as additives. Chevron refineries production systems are limited by thousands of operational constraints: Quality of crude oil comes with unique qualities and pricing; each refinery has a unique configuration and a different capability for handling various types of crude oil and manufacturing different product mixes. With market prices of crudes and products constantly in flux, determining the best way to operate the refineries is always a challenge. Chevron processes over two million barrels of crude oil per day. Any savings per barrel can be translated into millions of dollars annually.

Chevron decided to use the linear programming methodology to meet this challenge. Operations managers and business analysts at Chevron designed an in-house software tool called Petro. This distributive recursion-based linear programming-based software program is used by 25–30 people each day. Regardless of the complexity of the underlying models (400 to 500 variables and thousands of refinery-capability variables), the software is able to build new optimization cases within a few seconds and execute those cases in fractions of seconds. The linear programming-based program has generated about $600 million per year from operating the business and $400 million per year by enabling Chevron to make better decisions in capital allocation, for an estimated $10 billion over the past 30 years.

Introduction

Linear programming (LP) is considered to be one of the most commonly used techniques of management science in business organizations.[28] The popularity of this methodology can be explained by the nature of the LP models. LP models have two major components: an objective function and a set of constraints.[29] This structure allows for practitioners in every organizational system to strive to achieve goals under limited constraints. The following are some examples of goals and constraints to achieve them:

- A manager of a production line wants to minimize the total production and inventory cost (objective) while meeting the sales demands for its products (constraints).

- A financial analyst wants to maximize the return on investment (objective) with a limited budget for stocks and bonds (constraints).
- A marketing manager wants to maximize the customer lifetime value (objective) with a limited advertising budget (constraints).
- A logistics manager wants to minimize the total transportation costs for products and raw materials (objective) while meeting customer demands for such products (constraints).

Linear programming is a special type of mathematical programming. It is a mathematical programming approach where the objective function and constraints are assumed to be linear. The assumptions of linearity include: *proportionality* and *additivity*. The *proportionality* feature assumes that the contribution of product to the objective function and each resource to the constraint is proportional to the quantity of the product or the level of the resource. The *additivity* feature assumes that individual contribution of different products can be summed up to obtain an objective function or the individual usage of resources by each product can be summed up to obtain the total usage of the resource.

Algebraically, linear equations usually have constants and must have simple variables, such as x or y, but not x^2, or xy. Assume that a production line produces two products: *tables* and *chairs*. If the production cost for one table is \$50 and the line produces x units of them, then the total production cost for tables is $50x$. This is a *linear* relationship. If the production cost for one chair is \$30 and the line produces y units of them, then total production costs for chairs is $30y$. This is another linear function. The total production cost for both tables and chairs can then be expressed as an additive linear function: $50x + 30y$. If the production line were to increase its capacity by 10%, the production cost would increase by 10% as well. That is *proportionality* in a linear relationship.

Similarly, linear equations can be constructed to represent constraints. Suppose that the same production line is constrained by a limited number of labor hours available every week. If the production uses five hours of labor for a table and four hours of labor for a chair, then the total labor hours used for tables is $5x$ and the total labor hours for chairs is $4y$. The total use of labor can then be expressed as a linear function: $5x + 4y$. If the production line were to produce more tables and chairs, then the labor hours used would increase *proportionally* and *additively*.

LP models also assume that each parameter is known for sure (certainty) and that decision variables can take fractional as well as integer values (divisibility). When the constraints or the objective function cannot be expressed as linear equations, then other mathematical programming techniques can be used, such as nonlinear programming, integer programming, optimization heuristics, neural networks, or decision trees.

This chapter focuses on the understanding of fundamentals of mathematical programming in general and linear programming in particular. It demonstrates the importance of LP as a business analytics tool and discusses the potential use of LP to improve organizational performance. The model is explained graphically, and its solution is demonstrated graphically and via Microsoft Excel Solver.

LP Formulation

The formulation of the LP models is demonstrated with two examples. The first example is from a manufacturing environment, and the second example relates to a political marketing firm. Both examples can be represented graphically; the first example is a maximization problem, while the second example is a minimization problem. These examples are simple, but they will lay the foundation for the more complicated LP models discussed in the next chapters.

Example 1: Rolls Bakery Production Runs

Rolls Bakery produces two different types of products: dinner roll cases (DRC) and sandwich roll cases (SRC). The bakery has a total of 150 machine hours available among its baking lines during a typical production week. Each of the two products is produced in lots of 1,000 cases at a time. Each product has a different wholesale price, processing time, cost of raw materials, and weekly market demand. These values are shown in Table 2-1. The selling price for each product case is $0.75 for a DRC and $0.65 for a SRC. Because a production lot consists of 1,000 cases, the price per lot is $750 and $650, respectively, as shown in column 2 of Table 2-1. Each production lot of DRC requires 10 hours of processing time and each production lot of SRC requires 15 hours of processing time (column 3). The cost of raw materials is also calculated per lot (column 4), and it is $250 and $200, respectively. The labor cost per hour is $10.

Column 5 can be used to calculate the net profit per each lot. The net profit per each DRC lot can be calculated as follows: The price per one lot of DRC is $750 (column 2). The cost per lot is calculated by adding the cost of raw materials ($250) and labor cost ($10 × 10 hours = $100). As such, the net profit for a lot of DRCs is: net profit = $750 – ($250 – $100) = $400. Similarly, the net profit for a production lot of SRCs is: net profit = $650 – ($200 – $150) = $300. Both of these profits are shown in column 5.

Finally, the contractual agreements with the retailers require that Rolls Bakery produce no fewer than 3,000 cases of DRC and no fewer than 4,000 cases of SRC, as indicated in column 6. Because each production lot can prepare 1,000 cases, the weekly demand can be met by producing three lots of DRCs and four lots of SRCs (column 7). Rolls Bakery wants to determine how many production lots should run every week from each product to maximize the total net profit while meeting the weekly demand and not exceeding the available machine hours.

Table 2-1 Production Requirements for Rolls Bakery

Product	(1) Wholesale Price per Case	(2) Wholesale Price per Lot	(3) Processing Time (in Hours) per Lot	(4) Cost of Raw Materials per Lot	(5) Net Profit per Lot	(6) Demand for Cases	(7) Demand for Production Lots
Dinner roll case (DRC)	$0.75	$750	10	$250	$400	3,000	3
Sandwich roll case (SRC)	$0.65	$650	15	$200	$300	4,000	4

Formulation Steps

This section demonstrates the formulation of LP models with a four-step process: define decision variables, formulate the objective function, identify the set of constraints, and identify the set of non-negativity constraints. These steps are demonstrated with the example of Rolls Bakery.

Step 1: Define Decision Variables

Decision variables are those unknown quantities that the decision maker must identify in the process of solving the LP model. They are usually denoted as $x_1, x_2, x_3, \ldots x_n$. Defining the decision variables correctly is an important step in the LP formulation because these variables will set the stage for the rest of the formulation steps. As a practical matter, the decision maker can look for the decision variables by asking the following question:

What is the set of variables for which a value must be determined?

Usually the answer to this question is found in the problem description. For example, a manager must determine *how many units of each product* must be produced when trying to minimize the total production and inventory cost while meeting the sales demands for its products. A financial analyst must determine *how many units of each security* must be purchased when trying to maximize the return on investment within a limited budget. A marketing manager must determine *how many advertisements for each campaign* must be released when trying to maximize the customer responses. Finally, a logistics manager must decide *how many trucks must be assigned from each city to a destination city* when trying to minimize the total transportation costs for products and raw materials. In all four examples, *the number of products to be produced, the amount of securities to be purchased, the number of advertisement to be released,* and *the number of trucks to be assigned* serve as decision variables in the respective LP models.

Consider the Rolls Bakery problem. Its description states specifically that the bakery "wants to determine how many production lots from each product should run every week to maximize the total net profit while meeting the weekly demand and not exceeding the available machine hours." So the number of DRC lots and the number of SRC lots should be considered as decision variables. It is important to also include a time horizon when defining the decision variables. This sets the time framework for the rest of the equations in the LP model. In the example, the LP model represents a weekly production run. As such, the decision variables can be defined as follows:

x_1 = number of DRC lots to be produced during the week

x_2 = number of SRC lots to be produced during the week

Step 2: Formulate the Objective Function

The definition of *decision variables* can be used to formulate the equation representing the objective function. This step in itself can be divided into several smaller substeps:

1. **Define whether the goal is to maximize or minimize the objective function.**

   Every linear programming model has either a maximization or a minimization objective function. The modeler will use a maximization objective function when trying to optimize profit, production levels, customer satisfactions, revenue, and so on. The decision maker will use a minimization objective function when trying to optimize cost, processing times, production waste, budget, and so on. The goal of Rolls Bakery is to *maximize* profits; as such, the Rolls Bakery LP model is a maximization model.

2. **Identify contribution coefficients.**

   The next step in formulating the objective function is to identify the *contribution coefficients*. A contribution coefficient c_j represents the impact of one unit of the decision variable x_j in the objective function. Because the goal of Rolls Bakery is to maximize the total profit of the production, and because the decision variables indicate the number of lots to be produced during the week, the net profit from column contribution coefficients are shown in column 5 of Table 2-1. These values are 400 and 300, respectively, and will serve as the contribution coefficients.

3. Create the equation for the objective function.

The product $c_j x_j$ represents the amount of contribution of decision variable j in the objective function. Adding all the products together calculates the overall contribution in the form of objective function. In the case of Rolls Bakery model, the objective function is:

$$\text{Max } Z = c_1 x_1 + c_2 x_2 = 400 x_1 + 300\, x_2 \qquad (2.1)$$

where Z represents the total net profit contributions if the bakery decides to make x_1 a lot of DRCs and x_2 a lot of SRCs.

Step 3: Identify the Set of Constraints

Linear programming models will seek to maximize or minimize the objective function subject to a set of constraints. This step is rather daunting when formulating large LP models. The next chapter discusses these complexities. The Rolls Bakery example has only two major constraints: The production levels must meet or exceed weekly demand, and the bakery should not exceed the available machine hours.

Each constraint has two sides, the left-hand side and the right-hand side. Equality or inequality signs are used to enforce limits imposed by these constraints. In the Rolls Bakery case, the amount of machine hours available during the week is 150 hours. If one lot of DRCs takes ten hours of processing, x_1 lots of DRC require a total of $10 x_1$ hours. Similarly, x_2 lots of SRCs require a total of $15 x_2$ hours. The total hours required to produce these rolls should not exceed 150 available hours. As a result, the first constraint of the LP model can be written as:

$$10 x_1 + 15 x_2 \le 150 \qquad (2.2)$$

The next two constraints for Rolls Bakery are the need to meet the weekly demand for each type of roll. As such,

$$x_1 \ge 3 \text{ and } x_2 \ge 4 \qquad (2.3)$$

Step 4: Identify the Set of Non-Negativity Constraints

This is a rather formal step but very important nonetheless. Can Rolls Bakery produce a negative number of rolls? The answer is obviously no. However, the model must formally state this. As a result, the LP model is forced to generate non-negativity values by adding the following constraints:

$$x_1 \geq 0 \text{ and } x_2 \geq 0 \tag{2.4}$$

Overall Formulation

After completing these four steps, the LP model for Rolls Bakery results in the following formulation:

$$\text{Max } Z = 400x_1 + 300x_2$$

subject to:

$$10x_1 + 15x_2 \leq 150 \quad \text{(constraint 1)}$$
$$x_1 \geq 3 \quad \text{(constraint 2)}$$
$$x_2 \geq 4 \quad \text{(constraint 3)}$$
$$x_1, x_2 \geq 0 \quad \text{(non-negativity constraints)}$$

Formulating LP models is the first step toward achieving optimal solutions. The second step, solution methodology, is presented later in this chapter. But, first, another model formulation, a minimization model with two decision variables, is discussed.

Example 2: Political Advertisement Agency

Political Communications (PoliCom) is a marketing firm that specializes in political campaigns during elections. In the current election, the company is working for a gubernatorial candidate. The campaign has a goal to reach one million voters every week until election day. It is estimated that each advertisement has a nonmarginal impact. That is, the impact in the voting results is the same whether an advertisement reaches one voter five times or reaches five voters only once. Based on the data from past races, TV spots are considered more effective than radio campaigns. For example, on an average, each TV advertisement can reach 25,000 viewers and a radio advertisement can reach 12,500 listeners. The cost per each TV spot is $1,200 and the cost for each radio spot is $400. In addition to voter impact, TV campaigns also generate greater voter turnout and donations. As a result, the campaign has decided to purchase a minimum of 30 TV spots and no more than 60 radio spots every week. The goal is to identify how many TV and radio spots must be purchased every week so the campaign can minimize the overall advertisement cost.

Formulation Steps

The suggested LP formulation steps will be followed in this example as well.

Step 1: Define Decision Variables

The description of the PoliCom example states, "The goal is to identify how many TV and radio spots must be purchased every week." As such, the decision variables can be defined as:

x_1 = number of TV spots to be purchased during a week

x_2 = number of radio spots to be purchased during a week

Step 2: Formulate the Objective Function

The goal of PoliCom is to *minimize* the cost. In this example, the *contribution coefficients* are respectively 1,200 and 400. As such, the objective function can be stated as:

$$\text{Min } Z = 1200x_1 + 400x_2$$

where Z represents the total weekly costs if the campaign decides to purchase x_1 TV spots and x_2 radio spots every week.

Step 3: Identify the Set of Constraints

The number of voters reached by each TV ad is 25,000. If the campaign runs x_1 spots, then it is expected to reach $25,000x_1$ voters.

Similarly, it will reach $12,500x_2$ listeners. The total number of voters that must be reached is 1,000,000. So the first constraint of the problem can be stated as:

$$25,000x_1 + 12,500x_2 \geq 1,000,000$$

The next set of constraints for the campaign is the need to purchase at least 30 TV spots and the need not to exceed 60 radio spots every week. As such,

$$x_1 \geq 30 \text{ and } x_2 \leq 60$$

Step 4: Identify the Set of Non-Negativity Constraints

The following two constraints ensure that the model will not provide a negative value for the number of TV or radio spots:

$$x_1 \geq 0 \text{ and } x_2 \geq 0$$

Overall Formulation

The LP formulation of the advertisement campaign is summarized as:

$$\text{Min } Z = 1200x_1 + 400x_2$$

subject to:

$25,000x_1 + 12,500x_2 \geq 1,000,000$	(voter reach constraint)
$x_1 \geq 30$	(weekly TV spots constraint)
$x_2 \leq 60$	(weekly radio spots constraint)
$x_1, x_2 \geq 0$	(non-negativity constraints)

The next section demonstrates how to solve the previous LP models graphically. A graphical approach to solving LP models provides a visual and intuitive understanding of LP models and their solution processes. It visually demonstrates the feasibility area and the attempts of the objective function to reach an optimal solution.

Solving LP Models: A Graphical Approach

The following four steps are suggested to solve LP models graphically:

Step 1: Graph the area of feasible solutions.

Step 2: Graph the objective function.

Step 3: Find the coordinates for the point of optimal solution.

Step 4: Find the value of the objective function at the optimal solution.

Solution for Rolls Bakery Production Run

The previous steps are demonstrated for the Rolls Bakery problem. Here is the LP formulation as shown in the previous section:

$$\text{Max } Z = 400x_1 + 300x_2$$

subject to:

$10x_1 + 15x_2 \leq 150$	(constraint 1)
$x_1 \geq 3$	(constraint 2)
$x_2 \geq 4$	(constraint 3)
$x_1, x_2 \geq 0$	(non-negativity constraints)

Step 1: Graph the Area of Feasible Solutions

The Rolls Bakery problem contains two decision variables, three regular constraints, and two non-negativity constraints. Plotting the inequality equations of each constraint in a Cartesian coordinate system defines the area where all the points satisfy the set of constraints. The graphical area that satisfies all the constraints of the LP model is called the area of feasible solution. Figure 2-1 demonstrates the process of defining the area of feasible solution as the constraints are gradually added. The non-negativity constraints require that both decision variables (x_1 and x_2) remain positive, so the area of feasible solution always remains in the upper-right side of the Cartesian axes where all values for x_1 and x_2 are positive, as shown in Figure 2-1a.

Constraint 1 is graphed by picking two different values for one of the variables and solving for the corresponding values for the other variable. These values are then plotted as two points on the graph and the straight line connecting these two points is the line of constraint 1.

For example, if:

$x_1 = 0$ then $15x_2 = 150$, which results in $x_2 = 10$

Similarly, when

$x_2 = 0$ then $10x_1 = 150$, which results in $x_1 = 15$

The above values for x_1 and x_2 indicate that the straight line for constraint 1 will go through two points: $(0, 10)$ and $(15, 0)$. Because the left-hand side value (LHV) of the equation for constraint 1 is less than or equal to the right-hand side value (RHV) value, the shaded area below to the left of the line represents this constraint graphically. However, you should not assume that the area below or to the left of the line should always be shaded for less than or equal constraints. The best approach is to pick a point—for example, $(0, 0)$—and then substitute its coordinates into the inequality.

If the resulting statement is true, shade that side of the line; if not, shade the other side. Because $(0, 0)$ is clearly below the plotted line, and because substituting $x_1 = 0$ and $x_2 = 0$ into the constraint gives $0 < 150$ (which is a true statement), then the side of the line that includes the point $(0, 0)$ must be shaded.

The combination of two non-negativity constraints and constraint 1 limits the area of feasible solution to the ABC triangle shown in Figure 2-1a. Constraint 2 $(x_1 \geq 3)$ limits the area of feasible solution further, as shown in Figure 2-1b. Similarly, adding constraint 3 limits the area to the HIJ triangle, as shown in Figure 2-1c.

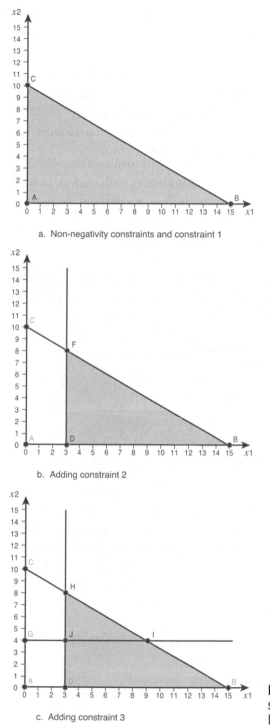

a. Non-negativity constraints and constraint 1

b. Adding constraint 2

c. Adding constraint 3

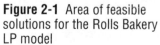

Figure 2-1 Area of feasible solutions for the Rolls Bakery LP model

Step 2: Graph the Objective Function

The objective function can be graphed in a similar way to the constraints. The only difference is that the equation representing the objective function has no right-hand side value. If an RHV value is arbitrarily assigned, then the line of the objective function can be graphed by determining two points, as explained in the previous section. For easy calculations, a value of zero can be assigned to the objective function, which forces its line to pass through the origin of the Cartesian space.

When assigning a value of zero in the objective function, the equation becomes:

$$Z = 400x_1 + 300x_2 = 0$$

If x_1 is equal to 0, then $300x_2$ is equal to 0, which leads to x_2 also becoming 0, as such the point $(0, 0)$. Another point is needed to determine the straight line: If $x_2 = 4$, then:

$400x_1 + 1{,}200 = 0$, which leads to $400x_1 = -1{,}200$ and $x_1 = -3$

So, the objective function must also go through point $(-3, 4)$. The objective function line, connecting the points $(0, 0)$ and $(-3, 4)$, is shown in red in Figure 2-2.

Step 3: Find the Coordinates for the Point of Optimal Solution

The goal is to determine the largest value of the objective function (for maximization problems) while the line is still touching the area of feasible solution. The optimal solution occurs at one of the corners of the feasible region. In the case of maximization problems, the line is moved upward parallel to itself (see the arrows in Figure 2-2) until the last point of contact with the area of feasible solution. In the case of minimization problems, the line is moved upward parallel to itself until the first point of contact with the area of feasible solution. As shown in Figure 2-2, the last point of contact of the objective function with the area of feasible solution is I (9, 4). That means that the optimal values for the number of x_1 and x_2 are: $x_1 = 9$ and $x_2 = 4$.

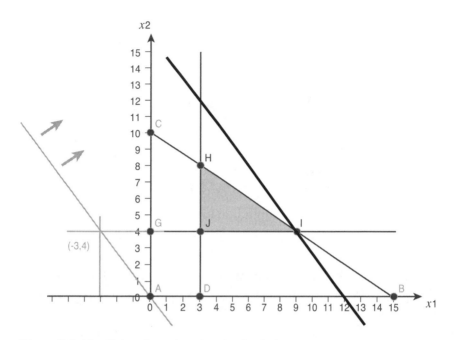

Figure 2-2 Identifying the point of optimal solution

Constraints 1 (machine hours) and 3 (demand for SRC lots) are called binding constraints. Binding constraints are represented by the lines that intersect at the point of the optimal solution. Constraint 2 (demand for DRC lots), to the contrary, is not a binding constraint. Resources represented by a binding constraint are fully used by the optimal solution. Because the optimal solution consists of nine DRC lots and four SRC lots, and because one DRC lot requires 10 hours of machine time and one SRC lot requires 15 hours of machine time, the overall time used by the optimal solution is $10 \times 9 + 15 \times 4 = 150$, which is the maximum amount of machine time available at Rolls Bakery. There is no slack in a binding constraint: The amount of used resources is equal to the amount of resources available. Constraint 3, which reflects the requirement that the bakery must produce at least four lots of SRCs is also completely satisfied because the optimal solution requires the production of exactly four lots ($x_2 = 4$). There is no slack here either. Constraint 2, however, is not a binding constraint and does have slack. Because it reflects the requirements that the bakery must make at least three lots of DRCs, and because the optimal solution indicates in fact a production of nine lots, the slack is six lots of DRCs.

Step 4: Find the Value of the Objective Function at the Point of Optimal Solution

The values of x_1 and x_2 found in the previous step are substituted into the objective function to find the value of the objective function at the optimal solution. In the Rolls Bakery model, the maximum value of the objective function is:

$$\text{Maximum } Z = 400 \times 9 + 300 \times 4 = 4{,}800$$

As shown graphically, to achieve a weekly profit of $4,800, Rolls Bakery must run nine production lots for the DRC and four production lots for the SRC.

Solution for the PoliCom Campaign

The LP model for PoliCom is summarized as:

$$\text{Min } Z = 1200x_1 + 400x_2$$

subject to:

$25{,}000x_1 + 12{,}500x_2 \geq 1{,}000{,}000$ (voter reach constraint)

$x_1 \geq 30$ (weekly TV spots constraint)

$x_2 \leq 60$ (weekly radio spots constraint)

$x_1, x_2 \geq 0$ (non-negativity constraints)

Step 1: Graph the Area of Feasible Solutions

The non-negativity constraints require that both decision variables remain positive, so the area of feasible solution always remains in the upper-right side of the Cartesian axes where all values for x_1 and x_2 are positive, as shown in Figure 2-3.

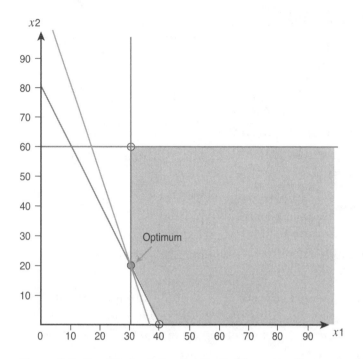

Figure 2-3 Graphical solution for the PoliCom advertisement problem

The first constraint $25{,}000x_1 + 12{,}500x_2 \geq 1{,}000{,}000$ can be graphed as follows:

Assign $x_1 = 0$; this results in $12500x_2 = 1{,}000{,}000$ which makes $x_2 = 80$

Assign $x_2 = 0$; this results in $25{,}000x_1 = 1{,}000{,}000$ which makes $x_1 = 40$

Thus, constraint 1 will go through two points: $(0, 80)$ and $(40, 0)$. Because the LHV of the equation for constraint 1 is greater than or equal to the RHV, the area of feasible solution will be on the upper-right side of the line. Constraint 2 ($x_1 \geq 30$) is represented by the vertical line that goes through point ($x_1 = 30$, $x_2 = 0$). Constraint 3 ($x_2 \leq 60$) is represented by the horizontal line that goes through point ($x_1 = 0$, $x_2 = 60$). The shaded area in Figure 2-3 shows the feasible solution. Every point in this area satisfies all three constraints. The objective function is used to find the point where the value of objective function (cost of campaign) is minimized.

Step 2: Graph the Objective Function

Assigning a value of zero to the objective function ($Z = 0$) leads to the following equation:

$$1200x_1 + 400x_2 = 0$$

One point in this line is the origin: $x_1 = 0$ then $400x_2 = 0$ and $x_2 = 0$

Another point can be identified by simply assigning a value to any of the decision variables. For example:

if $x_2 = 30$, then $1200x_1 + 400 \times 30 = 0$, which leads to $1200x_1 + 12{,}000 = 0$, thus $x_1 = -10$

As a result, the objective function line will connect the following two points: $(0, 0)$ and $(-10, 30)$.

A more practical way to create a starting line for the objective function is to equate the value of Z to the product of the contribution coefficients or to a common factor of them. For example, if $Z = 48{,}000$, then:

$x_1 = 0$, then $400x_2 = 48{,}000$ and $x_2 = 120$

$x_2 = 0$, then $1200x_1 = 48{,}000$, which leads to $x_1 = 40$

The objective function line will now connect the following two points: $(0, 120)$ and $(40, 0)$.

The set of lines representing the objective functions at different values for Z are parallel to each other. The line representing the objective function is known as the iso-profit (or in this, case iso-cost) line because all points in the line indicate a combination of coordinates, which result in an equal level of profit or cost.

Step 3: Find the Coordinates for the Point of Optimal Solution

The goal is to move the line parallel to itself until it reaches the lowest point of the feasible area. In the example under consideration, the optimal solution occurs at one of the corners of the feasible region. As shown in Figure 2-3, the last point of contact (when moving downward) of the objective function with the area of feasible solution is

(30, 20). That means that the optimal values for the number of TV spots and radio spots are: $x_1 = 30$ and $x_2 = 20$.

Step 4: Find the Value of the Objective Function at the Point of Optimal Solution

The values of x_1 and x_2 found in the previous step are substituted into the objective function to find the value of the objective function at the optimal solution. The minimum value of the objective function is: $Z = 1{,}200 \times 30 + 400 \times 20 = 44{,}000$. As shown graphically, to minimize the advertisement cost, PoliCom needs to run 30 TV spots and 20 radio spots on a weekly basis. This ensures that that the campaign reaches at least one million users and limits the number of radio spots to 60 per week and ensures that the company offers no fewer than 30 TV spots.

Possible Outcome Solutions to LP Model

As with many other mathematical models, solving an LP problem may sometimes result in multiple solutions, no solutions, or even unbounded solutions. The graphical approach to LP is an effective way to intuitively explain such situations.

Multiple Solutions

As discussed earlier in this chapter, the optimal solution of the LP model is found at the extreme point of the feasible area. This approach assumes that the slope of the iso-profit/iso-cost line of the objective function is different from the slope of the line representing any of the binding constraints. However, when the objective function has the same slope as a binding constraint, then the last point of contact of the objective function with the feasible area is not a single point. In that case, the LP model has an infinite number of solutions.

Assume that the contribution coefficient for the DRCs in the Rolls Bakery example has changed from $400 to $200. The new objective function will be:

$$\text{Max } Z = 200x_1 + 300x_2$$

The line representing the new objective function now has the same slope (200/300) as the first constraint (10/15). As a result, this line is parallel to the binding constraint, as shown in Figure 2-4.

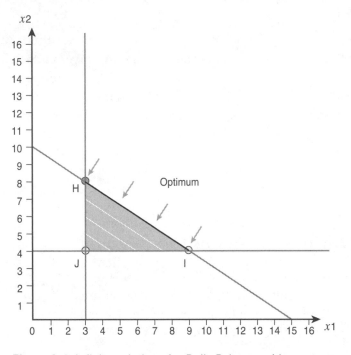

Figure 2-4 Infinite solutions for Rolls Bakery problem

A linear programming model with infinite solutions is generally good news. It offers many equally optimal options, and as a result the decision maker can choose the alternative that makes the best sense considering the other potential production requirements.

No Solutions

The graphical approach assumes that there exists an area of feasible solution. However, when too many restrictions have been placed in the model, the constraints may not have a common region. Assume that management has added a new constraint in the original Rolls Bakery problem:

Rolls Bakery is supplied by a limited amount of flour by a local supplier. The weekly supply is 1,000 lbs. and each case of rolls requires 0.2 lb., which results in 200 lbs. (0.2 × 1,000) per lot.

A fourth constraint is added to the original LP model, as shown here:

$$\text{Max } Z = 400x_1 + 300x_2 \text{ (objective function)}$$

subject to:

$10x_1 + 15x_2 \leq 150$	(constraint 1)
$x_1 \geq 3$	(constraint 2)
$x_2 \geq 4$	(constraint 3)
$200x_1 + 200x_2 \leq 1,000$	(constraint 4—flour supply)
$x_1, x_2 \geq 0$	(non-negativity constraints)

The new constraint is graphically represented by the area below line OP in Figure 2-5. As seen, the new region does not overlap with the original area HIJ. As such, no solution can be derived in this model.

Linear programming models with infeasible solutions often arise in practice because management expectations are too high. How should a data analyst interpret infeasibility? First, the analyst should derive that given the specific resources, it is not possible to make the required amount of products. In the Rolls Bakery example, considering the limited supply of flour, it is impossible to run nine lots of DRCs and four lots of SRCs. Second, the analyst should calculate what additional resources are needed to make the problem feasible. Because the minimum requirement is to run seven lots (three DRCs and four SRCs as indicated by constraints 2 and 3), then at least 7 × 200 = 1,400 lbs. of flour must be purchased every week. The bakery should consider buying from a second supplier or increasing the quantity from the first supplier.

It is important to realize that *infeasible solution* is indeed a solution. It helps determine whether management plans are feasible. Understanding the reasons for the infeasibility can lead to corrective actions and higher productivity.

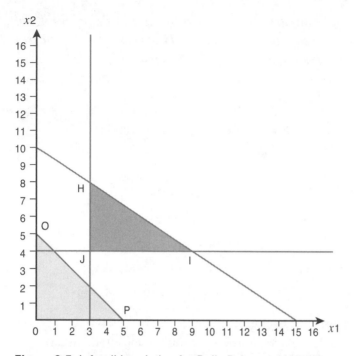

Figure 2-5 Infeasible solution for Rolls Bakery problem

Unbounded Solutions

In the Rolls Bakery case, the number of machine hours available during the week is limited to 150 hours. If this assumption is removed, then the original Rolls Bakery problem will become unbounded, as shown in Figure 2-6. Because there is no upper boundary in the machine hours, the solution to the problem can be as large as possible. From a practical perspective, the unbounded solution occurs when there is a missing constraint. A change in the objective function may result in a bounded solution. For example, in the bakery problem, changing the objective function to a minimization problem (in case the management wants to minimize labor costs) will transform the problem into a bounded problem.

The visual demonstration of graphical solutions for LP models with two decision variables can be helpful to better understand the structure and potential behavior of even larger problems with many decision variables and many constraints. The larger LP models cannot be graphed in a two-dimensional Cartesian space. Other methods such as the Simplex method can be used in such cases. Today, there are computer programs that can be used to enter LP formulations, solve them, and offer answer reports. The following sections demonstrate using Solver for the Rolls Bakery production and PoliCom advertising problems.

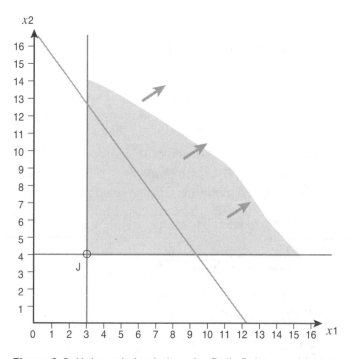

Figure 2-6 Unbounded solutions for Rolls Bakery problem

Solving LP Problems with Solver

Solver is a Microsoft Excel add-in that can be used to successfully solve LP models. The objective function and the constraints can be entered in the form of a template, and then Solver can be used by the decision maker to enforce those constraints, select the objective function, attempt a large number of solutions via multiple iterations, identify the best solution, and provide solution reports.

Using Solver for solving LP models has two advantages: accessibility and ease of use. Microsoft Excel is a popular spreadsheet software available as part of Microsoft Office. The availability of Excel permits the practitioners to have easy access to the Solver add-in, and allows them to share LP model templates and solutions. In addition, the practitioners are able to gain a better understanding of how to construct LP models utilizing Excel's convenient data entry, processing, and editing features. A complete tutorial in Solver is offered in Appendix B, "A Brief Tour of Solver."

Solving the Rolls Bakery Problem with Solver

The graphical solution of the Rolls Bakery problem with two decision variables was demonstrated earlier in this chapter. You are encouraged to review the graphical solution and compare the results with the Solver-based solution. This will allow for a better understanding of what Solver can do for LP models and create a good foundation when using larger LP models in the later chapters. From a practical perspective, using Solver for LP models is a three-stage process:

- Stage 1: Create an Excel template.
- Stage 2: Apply Solver.
- Stage 3: Interpret Solver solution.

Stage 1: Create an Excel Template

The Excel template of Solver is, in fact, a *what-if* model that allows the decision maker to calculate the value of the objective function and the usage of constraints for different input values in the decision variables. Figure 2-7 shows the template for the Rolls Bakery problem. The Excel file (*Ch2_RollsBakery*) for this example can be downloaded from the companion website at www.informit.com/title/9780133760354. The lower part of the figure shows the formulas "behind" the what-if template. (Formulas in Excel are displayed with the Ctrl + ~ combination).

The values shown in the template correspond to a trial solution of running one production lot for each type of roll. Cells B8 and B9 indicate the number of lots for DRC and SRC to run during the week. Cell C10 = C8 + C9 calculates the value of the objective function for the trial solution where C8 and C9 are calculated as the product between the number of lots and their respective net profit from cells F3 and F4. The same approach is used to calculate the number of machine hours used for the trial solution D10 = D8 + D9.

	A	B	C	D	E	F	G	H
1				Rolls Bakery Production Information				
2	Products	Wholesale Price per Case	Wholesale Price per Lot	Processing Time (in hours) per Lot	Cost of Raw Materials per	Net Profit per Lot	Demand for Cases	Demand for Production Lots
3	Dinner Roll Case (DRC)	$0.75	$750	10	$250	$400	3000	3
4	Sandwich Roll Case (SRC)	$0.65	$650	15	$200	$300	4000	4
5								
6			What-If Analysis					
7	Product	Number of Lots per Week	Net Profit Per Lot	Total Machine Time Used (in hours)	Right hand-side values			
8	Dinner Roll Case Lots	1	$400	10	3	Minimum DRC (Constraint 2)		
9	Sandwich Roll Case Lots	1	$300	15	4	Minimum SRC (Constraint 3)		
10		Total:	$700	25	150	Machine Hours (Constraint 1)		

	A	B	C	D	E	F	G	H
1				Rolls Bakery Production Information				
2	Products	Wholesale Price per Case	Wholesale Price per Lot	Processing Time (in hours) per Lot	Cost of Raw Materials per	Net Profit per Lot	Demand for Cases	Demand for Production Lots
3	Dinner Roll Case (DRC)	0.75	=B3*1000	10	250	=C3-D3*10-E3	3000	=G3/1000
4	Sandwich Roll Case (SRC)	0.65	=B4*1000	15	200	=C4-D4*10-E4	4000	=G4/1000
5								
6			What-If Analysis					
7	Product	Number of Lots per Week	Net Profit Per Lot	Total Machine Time Used (in hours)	Right hand-side values			
8	Dinner Roll Case Lots	1	=B8*F3	=B8*D3	=H3	Minimum DRC (Constraint 2)		
9	Sandwich Roll Case Lots	1	=B9*F4	=B9*D4	=H4	Minimum SRC (Constraint 3)		
10		Total:	=C8+C9	=D8+D9	150	Machine Hours (Constraint 1)		

Figure 2-7 Excel spreadsheet model for Rolls Bakery

Stage 2: Apply Solver

Once the LP template is built in Excel, Solver is used to enforce the constraints and the objective function. The Solver Parameters dialog box is shown in Figure 2-8. The following steps are the steps for entering the LP formulation into Solver:

1. Select the Data tab in the Ribbon.
2. Click Solver in the Analysis group.

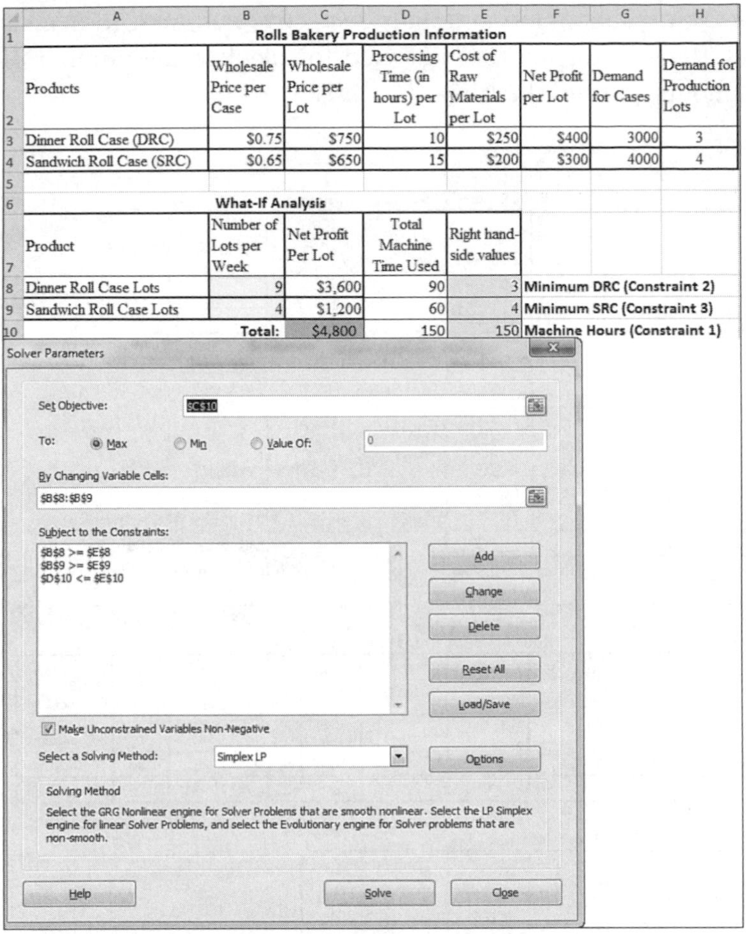

	A	B	C	D	E	F	G	H
1			Rolls Bakery Production Information					
2	Products	Wholesale Price per Case	Wholesale Price per Lot	Processing Time (in hours) per Lot	Cost of Raw Materials per Lot	Net Profit per Lot	Demand for Cases	Demand for Production Lots
3	Dinner Roll Case (DRC)	$0.75	$750	10	$250	$400	3000	3
4	Sandwich Roll Case (SRC)	$0.65	$650	15	$200	$300	4000	4
5								
6			What-If Analysis					
7	Product	Number of Lots per Week	Net Profit Per Lot	Total Machine Time Used	Right hand-side values			
8	Dinner Roll Case Lots	9	$3,600	90	3	Minimum DRC (Constraint 2)		
9	Sandwich Roll Case Lots	4	$1,200	60	4	Minimum SRC (Constraint 3)		
10		Total:	$4,800	150	150	Machine Hours (Constraint 1)		

Solver Parameters

Set Objective: $C$10

To: ◉ Max   ○ Min   ○ Value Of: 0

By Changing Variable Cells:
$B$8:$B$9

Subject to the Constraints:
$B$8 >= $E$8
$B$9 >= $E$9
$D$10 <= $E$10

Add
Change
Delete
Reset All
Load/Save

☑ Make Unconstrained Variables Non-Negative

Select a Solving Method: Simplex LP   Options

Solving Method
Select the GRG Nonlinear engine for Solver Problems that are smooth nonlinear. Select the LP Simplex engine for linear Solver Problems, and select the Evolutionary engine for Solver problems that are non-smooth.

Help   Solve   Close

Figure 2-8 Solver Parameters dialog box and solution

3. Solver will appear in a separate box.
   a. Enter C10 in the Set Objective field.
   b. Select Max (because the Rolls bakery is a maximization problem).
   c. Enter B8:B9 in the By Changing Variable Cells field.
4. Adding Constraints: Click the Add button.
   a. The Add Constraint dialog box appears.
      • Enter B8 in the left-hand box under Cell References.
      • Select >= from the drop-down list.
      • Enter E8 in the Constraint field.

  b. Repeat the process to enter the other two constraints, as shown in Figure 2-8.

  c. Click OK.

5. Select the Make Unconstrained Variables Non-Negative check box.

6. Choose Simplex LP from the Select a Solving Method drop-down list.

7. Click Solve.

8. When the Solver Results dialog box appears:

  a. Select Keep Solver Solution.

  b. In the Reports Section, select Answer Report.

  c. Click OK.

As shown, the solution is similar to the one achieved via the graphical approach: nine production runs for DRCs and four production runs for SRCs. The total profit of $4,800 is shown in cell C10.

Stage 3: Interpret Solver Solution

  The Answer Report is shown in Figure 2-9. It contains three parts: Objective Cell, Variable Cells, and Constraints:

- **Objective Cell**—This cell indicates the final solution of the total profit of $4,800. The original value shown in the figure is the value of the objective function for the initial trial solution.

- **Variable Cells**—Note the final value in the variable cells. It indicates the optimal solution for the decision variables, nine and four.

- **Constraints**—This section shows the left-hand side value for each constraint (under Cell Value), the direction of the constraint (under Formula), the Status, and the Slack. Notice the relationship between Status and Slack. Binding constraints always have a slack of 0 and not binding constraints have a positive value as slack.

In addition to the Answer Report, Solver also generates a Sensitivity Report and a Limits Report. The information provided in these two reports is discussed in the sensitivity analysis section of Chapter 3, "Business Analytics with Linear Programming."

14	Objective Cell (Max)					
15	Cell	Name		Original Value	Final Value	
16	$C$10	Total: Net Profit Per Lot		$700	$4,800	
17						
18						
19	Variable Cells					
20	Cell	Name		Original Value	Final Value	Integer
21	$B$8	Dinner Roll Case Lots Number of Lots per Week		1	9	Contin
22	$B$9	Sandwich Roll Case Lots Number of Lots per Week		1	4	Contin
23						
24						
25	Constraints					
26	Cell	Name	Cell Value	Formula	Status	Slack
27	$D$10	Total: Total Machine Time Used (in hours)	150	$D$10<=$E$10	Binding	0
28	$B$8	Dinner Roll Case Lots Number of Lots per Week	9	$B$8>=$E$8	Not Binding	6
29	$B$9	Sandwich Roll Case Lots Number of Lots per Week	4	$B$9>=$E$9	Binding	0

Figure 2-9 Solver Answer Report for Rolls Bakery problem

Solving the PoliCom Problem with Solver

The same approach can be followed to find the optimal number of TV and radio advertisements during the week, as shown in Figure 2-10. Note that in the Solver Parameters dialog box, the Min option is selected because this is a minimization problem. As the suggested steps are followed, the decision maker can identify the constraints, the objective function, the final values of the decision variables, the final value of the objective function, the binding constraints, and the not binding constraint with its respective slack. Figure 2-11 shows the Answer Report for the same problem. As shown, the report indicates the same solution as the one achieved via the graphical approach. Download the *Ch2_PoliCom* file from the companion website to further investigate the implementation of the three steps of applying Solver in the PoliCom advertising problem.

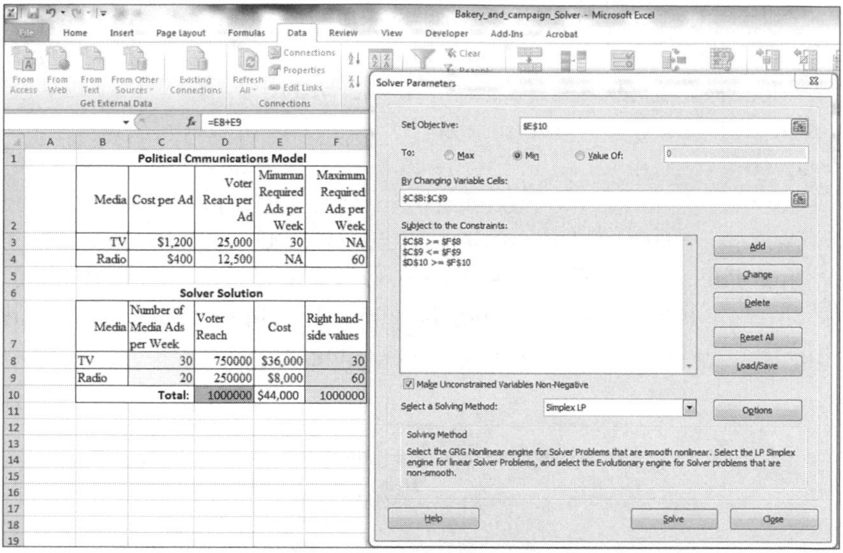

Figure 2-10 Solving PoliCom problem with Solver

13							
14	Objective Cell (Min)						
15	Cell	Name	Original Value	Final Value			
16	$E$10	Total: Cost	$1,600	$44,000			
17							
18							
19	Variable Cells						
20	Cell	Name	Original Value	Final Value	Integer		
21	$C$8	TV Number of Media Ads per Week	1	30	Contin		
22	$C$9	Radio Number of Media Ads per Week	1	20	Contin		
23							
24							
25	Constraints						
26	Cell	Name	Cell Value	Formula	Status	Slack	
27	$D$10	Total: Voter Reach	1000000	$D$10>=$F$10	Binding	0	
28	$C$8	TV Number of Media Ads per Week	30	$C$8>=$F$8	Binding	0	
29	$C$9	Radio Number of Media Ads per Week	20	$C$9<=$F$9	Not Binding	40	

Figure 2-11 Solver Answer Report for PoliCom advertising problem

Exploring Big Data with LP Models

Implementing LP models with Big Data input is an important, but difficult task. The chapter lays practical foundations for formulating and solving LP models by suggesting a step-by-step approach. The necessary steps to formulate an LP model are summarized in Figure 2-12. The examples used in this chapter are limited to LP models with two decision variables. However, the suggested methodology can be successfully used for large models as well. The decision maker must follow the steps outlined in Figure 2-12 to avoid the complexity of the LP models. It allows the modeler to implement a "model less, iterate more" approach to improve the efficiency of the optimization algorithms. The approach assumes that once a model is formulated, it can be used multiple times with newly calculated input parameters, and thus the model can be fine-tuned accordingly.

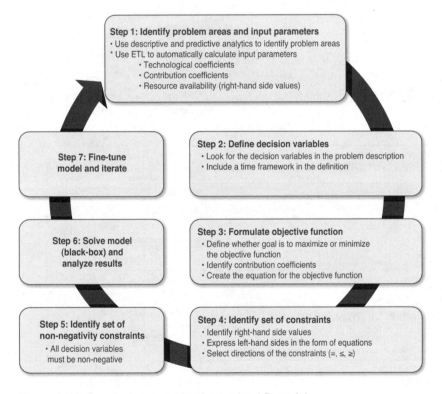

Figure 2-12 Practical steps to implementing LP models

Wrap Up

Figure 2-13 summarizes the steps for solving LP models graphically. The graphical solution to LP models may not be very practical considering the availability of more sophisticated programming tools. However, the graphical approach provides an excellent opportunity to the decision maker to understand the structure and behavior of LP models. This understanding is very important when the modeler deals with formulation and solution of large and complex LP models.

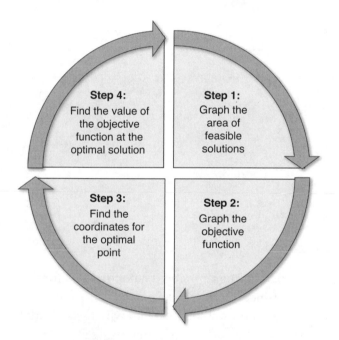

Figure 2-13 Graphical solution steps

Finally, Figure 2-14 summarizes three steps used to solve LP models with Solver. The same suggested Solver steps can be used for larger LP models, as illustrated in the next chapter. Appendix B also provides a detailed explanation of Solver as the main processing engine of optimization models throughout the book.

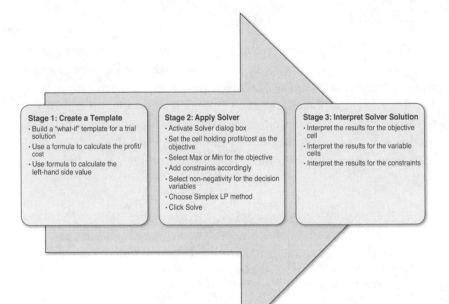

Figure 2-14 Solver solution steps

Review Questions

1. Why do you think linear programming is a popular tool for data scientists in the business world? Provide business situations where LP models can be used, and explain why these models are appropriate in the decision-making process.

2. Explain the concept of linearity. Provide examples that illustrate linear relationships among business entities. Provide examples that illustrate nonlinear relationships among business entities. Use additive and proportionality concepts to support your position.

3. Discuss challenges that a business analyst might face when formulating a linear programming model. For each challenge, discuss practical steps on how to overcome them.

4. Discuss the components of an LP model and the major steps for formulating an LP model. Why it is important to follow the steps carefully?

5. Defining the decision variables is a very important first step in the LP formulation. Explain why. What is a practical rule of thumb when deciding the set of decision variables? Why is it important to include a time component in the definition of decision variables?

6. What are the suggested steps to formulate an objective function? Discuss situations when the objective function should be maximized. Discuss situations when the objective function should be minimized.

7. What are the suggested steps in establishing a set of constraints? Discuss situations in which the left-hand side value is less than or equal to the right-hand side value. Discuss situations when the left-hand side value is greater than or equal to the right-hand side value. Discuss situations when the left-hand side value is exactly equal to the right-hand side value.

8. What is the difference between a binding and not binding constraint? If additional funding is available to purchase additional resources, should the company invest in resources that are represented by a binding or not binding constraint?

9. Can an LP model result in more than one optimal solution? If yes, explain the circumstances when this might happen and how a decision maker can select only one of the optimal solutions. If no, explain why not.

10. Can an LP model result in no solutions? If yes, explain the circumstances when this might happen and how a decision maker can avoid such cases. If no, explain why an LP should always result in one or more solutions.

11. Can an LP model result in unbounded solutions? If yes, explain the circumstances when this might happen and how a decision maker can avoid such cases. If no, explain why an LP should always be bounded.

Practice Problems

1. A computer company produces laptop and desktop computers. The Marketing Department projects that the expected demand for laptops will be at least 1,000, and for desktops it will be at least 800 per day. The production facility has a limited capacity of no more than 2,000 laptops and 1,700 desktops per day. The Sales Department indicates that contractual agreements of at most 2,000 computers per day must be satisfied. Each laptop computer generates $600 net profit and each desktop computer generates $300 net profit. The company wants to determine how many of each type should be made daily to maximize net profits.

   a. Formulate a linear programming model that represents the preceding business description.

   b. Solve the problems graphically and indicate:

      • What is the optimal number of laptop and desktop computers to be made each day?

      • What is the value of the objective function (total net profit) for the solution?

      • Identify binding and not binding constraints for the optimal solution.

   c. Solve the same problem using Excel's Solver.

2. You run an animal shelter facility and are asked to purchase food supply for the next month. You normally purchase two types of food, Purina and Natural Choice. Purina costs $180 per bag. Each bag weighs 80 pounds and can feed as many as six animals per month. The Natural Choice bag costs $110, weighs 60 pounds, and can feed four animals per month. You have been given a maximum budget of $4,000 for this purchase, although you don't have to spend that much. The facility has room for no more than 2,000 pounds and can board an average of 100 animals per day during a typical month. How many bags of each type of food should you buy to minimize the purchasing cost?

a. Formulate and solve the problems graphically to determine:

- What is the optimal number of Purina and Natural Choice bags to be purchased each month?
- What is the value of the objective function (total purchasing cost) for the preceding solution?
- Identify binding and not binding constraints for the optimal solution.

b. Solve the same problem using Solver and compare the Solver solution to the graphical solution.

3. Consider the same animal shelter described in problem 2. Solve the problem to identify how many bags of each food to purchase to feed the maximum possible number of animals during the month within the limited amount of budget.

a. Solve the problems graphically and compare it with the graphical solution from problem 2 regarding:

- The optimal number of Purina and Natural Choice bags to be purchased each month
- The value of the objective function; that is, the total number of animals able to feed
- Changes that occurred in binding and not binding constraints when comparing the solutions of these two problems

b. Use Solver to confirm the accuracy of the graphical solution.

4. A toy manufacturing company makes two toys, trucks and cars. The manufacturing requirements for each toy production lot include the availability of raw materials and usage per unit of each material in each toy. Specifically, these requirements are shown in the following table. For example, for each production lot, truck toys require six pounds of plastic, ten hours of labor, and ten hours of machine time. A car toy lot requires eight pounds of plastic, eight hours of labor, and four hours of machine time. There are 72 pounds of plastic available, 80 hours of labor available, and 60 hours of machine time for each day. The profit for either toy is $1,000 per lot.

Raw Materials	Truck Toy	Car Toy	Available
Plastic	6 lb.	8 lb.	72 lb.
Labor hours	10 hrs.	8 hrs.	80 hrs.
Machine time	10 hrs.	4 hrs.	60 hrs.

Formulate a linear programming model that represents the preceding business description.

a. Solve the problems graphically and indicate:

- What is the optimal number of truck toy lots and car toy lots to be purchased each day?

- What is the value of the objective function (total profit) for the preceding solution?

- Identify binding and not binding constraints for the optimal solution.

b. Solve the same problem using Solver and compare it with the graphical solution.

5. Consider the same toy production problem described in problem 4. Solve the problem to identify how many lots of toy trucks and toy cars to produce to minimize the overall utilization of labor hours and achieve a profit of at least $9,000. Solve the problem graphically and compare it with the graphical solution from problem 4 regarding:

a. The optimal number of toy truck lots and toy car lots to be produced each day

b. The value of the objective function, that is, the total number of labor hours used

c. Changes occurred in binding and not binding constraints

d. Use Solver to confirm the accuracy of the graphical solution.

e. Resolve the toy manufacturing problem both graphically and via Solver after changing the profit constraint. The new requirement seeks to achieve a profit of at least $10,000. Does the new constraint change the nature of the solution? If yes, how?

6. A paint manufacturer produces two paint primer products: Glidden Professional Aquacrylic (GPA) and BEHR Premium Plus (BPP). The manufacturer currently sells more units and receives better customer reviews from GPA than the BPP. Over the last few months, GPA has received an average rating of 4.1 out of 5 and BPP has received an average rating of 3.1. However, BPP offers a larger profit margin. The following table shows the sales price and manufacturing cost for each primer product.

Pain Primer	Sales Price per Gallon	Cost per Gallon
GPA	$20	$15
BPP	$15	$5

The production facility wants to stay under a $30,000 monthly budget for making these two products. Due to contractual agreements, the facility must produce at least 1,500 gallons of GPA and 1,000 gallons of BPP for each month. The goal of the manufacturer is to maximize the net profit.

a. Formulate a linear programming model that represents the preceding business description. Solve the problems graphically. (*Hint:* Consider changing the measuring unit from *gallons* to *hundreds of gallons* or *thousands of gallons* so the graphical solution can better fit in the x-y axis.)

  - What is the optimal number of GPA and BPP to be produced each month?

  - What is the value of the objective function (total profit) for the solution?

  - Identify binding and not binding constraints for the optimal solution and explain why some constraints are not fully met.

b. The manufacturer is also interested in the product reviews and wants to provide a production mix that maximizes the total customer satisfaction. Assuming that the average ratings are offered for each purchased gallon, reformulate the

LP model and answer the same questions as in step a. Does the solution change? Why or why not?

c. The manufacturer now changes the goal: It wants to minimize the production cost. However, a $20,000 profit per month must be achieved. Solve the problem and investigate the Solver's Answer Report to identify:

- The optimal monthly production mix
- Binding and not binding constraints
- Does the solution change compared with step b? Why or why not?

7. You are the logistics manager for a building supply company. The company has warehouses in two cities, Atlanta, Georgia, and Memphis, Tennessee. The sales office receives orders from two customers, each requiring dry wall sheets, delivered in boxes, which contain 10 sheets per box. Customer A needs 10 boxes of dry wall sheets and Customer B needs 14 boxes. The warehouse in Atlanta has 16 boxes in stock; the Memphis warehouse has 9 boxes in stock. Delivery costs per each box from a warehouse to a customer are shown in the following table:

From/To	Customer A	Customer B	Boxes in Stock
Atlanta	$10	$12	16
Memphis	$8	$11	9
Customer needs	10	14	

As a logistics manager, you need to find out how many dry wall boxes should be shipped from each city to each customer to minimize transportation cost and meet customer needs. Specifically:

a. Represent the preceding business description with a linear programming model *with two decision variables.*

b. Solve the problems graphically and indicate:
- What is the optimal number of sheets to be shipped from Atlanta to Customer A?

- What is the optimal number of sheets to be shipped from Atlanta to Customer B?

- What is the optimal number of sheets to be shipped from Memphis to Customer A?

- What is the optimal number of sheets to be shipped from Memphis to Customer B?

8. You are the financial manager of a production company. An important part of your role is determining the most profitable allocation of the company end-of-year cash on hand. This year, you have estimated that the company will have $400,000 end-of-year cash on hand. You are considering investing this amount and must determine the best way to allocate these funds in stocks and bonds.

   The stock fund has a 9.5% rate of return, but is very volatile, two times more volatile than the market average. The bond fund has a 4.5% rate of return, but is less volatile, only 90% compared with the market average. Stocks are also more expensive than bonds. The expense ratio for stocks is 1.5% of total investment, whereas the expense ratio for bonds is 0.2% of the total investment. The goal is to keep the total fund expense under $4,000. The chief financial officer has instructed you to invest at least half of the budgeted dollars in bonds to balance the riskier stock investment. Using a linear programming model, determine how you should allocate the end-of-year cash on hand to maximize the company's return.

   a. Formulate a linear programming model that seeks to maximize the rate of return in spite of volatility of the investment mix. Solve the problems with Solver and indicate:

      - What is the dollar amount to invest in stocks and bonds at the end of the year?

      - What is the value of the objective function (total profit) for the preceding solution?

      - Identify binding and not binding constraints for the optimal solution.

b. Reformulate the linear programming model from step a with a new goal: Minimize the overall volatility of the investment mix while the overall annual return must be at least $30,000. Solve the problem with Solver and determine the amount of funds invested in stock and bonds.

9. A small chiropractic facility is looking to maximize its weekly profits by determining how many new patients and regular patients to schedule each week. The practice has two types of patient appointments: regular treatments and new patient consultations/treatments. Labor and cost associated with treatments are the same for existing and new patients, whereas the amount of time scheduled is different for the two types of appointments. A new patient visit typically takes one hour and is performed by both the doctor and a technician. A regular patient takes 30 minutes and usually is performed by a technician. Because there are six treatment rooms, patients can be scheduled simultaneously; however, the chiropractor can schedule for no more than 30 hours per week and each technician can work as many as 40 hours per week.

The rate for a new patient consultation/treatment is $160 and the rate for an existing patient is $80. The cost of treating a patient is $25 and is the same for both existing and new patients. Determining the optimal number of new and existing patients to be scheduled each week allows the chiropractic facility to predict when to accept new patients as the existing patients complete their last treatment.

Formulate a linear programming model to determine the optimal number of new and existing patients to be scheduled every week. The goal is to maximize the net profit (treatment rate minus cost of treatment). Solve the problem using Solver.

3

Business Analytics with Linear Programming

Chapter Objectives

- Emphasize the importance of linear programming (LP) as a prescriptive analytics tool
- Provide a general formulation to LP models using mathematical notations
- Demonstrate the use of Excel and Solver for solving LP models with a large number of decision variables and constraints
- Discuss the importance of data preparation techniques and the use of pivot tables to summarize data and refresh data
- Perform sensitivity analysis and understand the impact of the changes in the right-hand side values and contribution coefficients
- Discuss the challenges of implementing LP models in the era of Big Data

Prescriptive Analytics in Action: Nu-kote Minimizes Shipment Cost[1]

Nu-kote is the world's largest independent remanufacturer of paint primer for printers, copiers, and fax machines, with a global network of suppliers, production facilities, and customers. An industry leader in the development of color and monochrome toner, Nu-kote's facilities are located in various cities in the United States, China, Thailand, and Mexico. As the company became more involved in managing its global operations, the management team needed to better plan shipments of finished goods from plants to warehouses, vendors, and customers.

A team of data scientists from Nu-kote and an external consulting team from Vanderbilt University developed a spreadsheet linear programming (LP) model to address such shipment problems. The main goal of the team was to minimize the overall shipment cost subject to maximum-shipping-distance policies. Different versions of LP models were used to optimize supply chains with different warehouse configurations. The LP models were relatively large: between 5,000 and 9,700 variables and about 2,500 constraints. Solving such a large, complex LP model required a combination of spreadsheet modeling with visual basic for applications (VBA) programming code. It is estimated that Nu-kote saved approximately $1,000,000 in its first year, with much higher returns in the subsequent years.

Introduction

The previous chapter offered an introduction to linear programming models. LP models have an objective function and a set of constraints. This structure makes the LP approach very popular for business situations because managers normally seek to maximize or minimize certain business goals under several resource-driven

---

[1] LeBlanc, L.J., and Galbreth, M.R. "Implementing Large-Scale Optimization Models in Excel Using VBA." *Interfaces* 37, no. 4 (2007): 370–382.

limitations. In fact, LP models are considered to be one of the top ten most important algorithms of the last century.[30] A recent literature survey [31] of LP applications cites various business situations where LP models are used. Applications of optimization problems extend over many fields. LP models are successfully applied to production and shipping, asset allocation, mortgage-backed securities portfolios, or other similar problems.

In spite of the popularity of linear programming, the application of large LP models in real-world optimization problems has been a challenge mainly due to their computational difficulty. Decision makers have used software programs to deal with computational aspects of LP models. These software packages include Analytic Solver Platform, CMPL, CPLEX Optimization Studio, GAMS, GIPALS - Linear Programming Environment, GNU Linear Programming Kit, LINDO, LINGO, Risk Solver Platform, among others. These tools offer reliable solutions with calculation efficiency for large models. However, these programs, besides being somewhat costly, require that users spend a significant amount of time to learn them.

Spreadsheet modeling is a popular approach to LP modeling and solutions. It offers a template-based approach to LP formulations, simplifies the setup process, transforms operational data into LP input parameters, and saves time for decision makers when these templates are used repeatedly. This chapter explores the formulation of complex LP models with Microsoft Excel spreadsheets and solutions for these models using Solver. Special emphasis is placed on data collection and preparation, which is considered to be one of the challenges of applying LP models in real business environments.

The output analysis of LP models is probably the most important component of business analytics. In the era of Big Data, it is important to have good models that provide accurate results. However, it is more important to have practitioners who can interpret the results correctly and use those results for smart decisions that offer real business value.

General Formulation of LP Models

The general notation of the LP model formulation is:

$$Max \ (or \ Min) \ Z = \sum_{j=1}^{n} c_j x_j$$

subject to:

$$\sum_{j=1}^{n} a_{ij} x_j \begin{cases} \leq \\ \geq \\ = \end{cases} b_i \ \text{for all } i = 1, 2, 3, ..., m$$

$$x_1, x_2, ... x_n \geq 0$$

where n is the number of decision variables and m is the number of constraints.

The decision maker can seek to maximize or minimize the objective function Z. c_j represents the set of contribution coefficients and x_j represents the set of decision variables where $j = 1...n$. Also, b_i represents the set of i right-hand side values for each constraint where $i = 1...m$. The direction of the constraints can be less than or equal to, greater than or equal to, or even equal to. a_{ij} are known as *technological coefficients* and represent the usage of the constraint i per one unit of the decision variable j.

Formulating a Large LP Model

Real-life optimization models are more complex than the LP model with two decision variables. However, regardless of the number of decision variables or the number of constraints, management scientists and practitioners can still use the same four steps suggested in the previous chapter: define decision variables, formulate the objective function, identify the set of constraints, and identify a set of non-negativity constraints. However, considering the complexity of the large amount of input data into such problems, calculating parameters of the model must be considered as an additional and first step. Thus, here are the five suggested steps for large LP models:

1. Calculate model parameters.
2. Define decision variables.
3. Formulate the objective function.
4. Identify the set of constraints.
5. Identify the set of non-negativity constraints.

The following section demonstrates these steps using the Primer Manufacturer Inc. example.

Example: Primer Manufacturer Inc. Production Mix

Primer Manufacturer Inc. (PMI) produces 48 different paint primers and distributes them to its clients on a weekly basis. The operations manager at the company wants to determine how many gallons of each primer to produce each week so she can maximize revenue under a limited number of available machine hours of 8,500 per week. Contractual agreements with several stores impose a minimum production level of 200 gallons a month for each primer and a maximum production level of 1,200 gallons for each primer. In addition, the factory has a limited budget for raw materials of $90,000.

The operational data at PMI is continuously recorded, and the manager has decided to build a model that uses records from the last 12 weeks. An Excel file named *ch3_PaintPrimer* has been created for this demonstration and can be downloaded from the book's companion website at www.informit.com/title/9780133760354. The first worksheet named operational data (partially shown in Figure 3-1) indicates the record for each primer during the last 12 weeks. A total of 576 records (12 weeks × 48 products) is used to calculate the average processing time and the average cost of raw materials.

	A	B	C	D
		Production	Processing Time (in	Raw Materials
1	Primer Product Number	Week	hours) per Gallon	Cost
2	KILZ-163TL	12	0.20	$4.11
3	KILZ2-166TL	12	0.63	$3.82
4	ZINSSER-182TLD	12	0.13	$1.94
5	GLIDEN-184TL	12	0.67	$4.50
6	GLIDEN PRO-186L	12	0.75	$2.61
7	KILZ PRO-192LT	12	0.12	$3.47
8	KILZ AERO-206TL	12	0.11	$4.53
9	GLIDEN PRO + SEALER-207TL	12	0.34	$2.54
10	KILZ PREMIUM-234TL	12	0.24	$2.75
11	BEHR-258TL	12	0.31	$2.95
12	BEHR PRO-268TL	12	0.71	$1.71
13	BEHR AERO-269LC	12	0.27	$3.90
14	BEHR PRO + SEALER-327TL	12	0.06	$4.04
15	ZINSSER PRO-69TL	12	0.11	$4.00
16	ZINSSER SEALER-82C	12	0.01	$1.98
17	ZINSSER PREMIUM-86L	12	0.17	$1.14
18	GILDEN PREMIUM-86LTORG	12	0.58	$3.60
19	KILZ-86TL	12	0.16	$1.22
20	KILZ2-B115	12	0.50	$4.52
21	ZINSSER-B126F	12	0.09	$2.97
22	GLIDEN-B127	12	0.85	$4.65
23	GLIDEN PRO-B128	12	0.66	$3.59
24	KILZ PRO-B129	12	0.14	$2.92

Figure 3-1   A subset of operational records

Step 1: Calculate Model Parameters

A pivot table is an excellent data summarization tool found in Excel and other business intelligence software, which can be used to calculate LP input parameters. This function can automatically sort, count, total, or average data stored in a spreadsheet. The results are displayed in a second spreadsheet, called a *pivot table*. Managers and practitioners use pivot tables to calculate average values of operational data, which in turn can be used as input parameters for the LP model. The Refresh option (Data, Refresh All) in the pivot table allows the decision maker to always update the model parameters and allows for the LP model to always use the latest operational data. These parameters can be calculated by dragging and dropping fields graphically.

The features and capabilities of the pivot table and how a pivot table can be used to calculate contribution coefficients and technological coefficients are shown below:

1. Select the relevant columns for data processing.

2. Using the Insert menu, click the Insert tab and click Pivot Table, as shown in Figure 3-2.

3. In the Create PivotTable dialog box, select New Worksheet in the Choose Where You Want the PivotTable Report to Be Placed section to place the pivot table in a new worksheet (Figure 3-2, bottom).

4. Click OK. The pivot table is now placed on a new worksheet as shown in Figure 3-3.

5. Choose fields form the PivotTable Field List dialog box to start building the pivot table report.

The Primer Product Numbers field is added to the pivot table report by clicking its check box or by dragging the field to the Row Labels section of the PivotTable Field List dialog box. Processing Time (in hours) per Gallon and Raw Materials Cost are placed in the Values section. By default, the Count function is applied to the fields placed in the Values section. However, you can select the desired function by clicking in the drop-down menu of the field and selecting the Value Field Settings options. In our example, the default Count function is changed to the Average function for both fields: Processing Time (in hours) per Gallon and Raw Materials Cost.

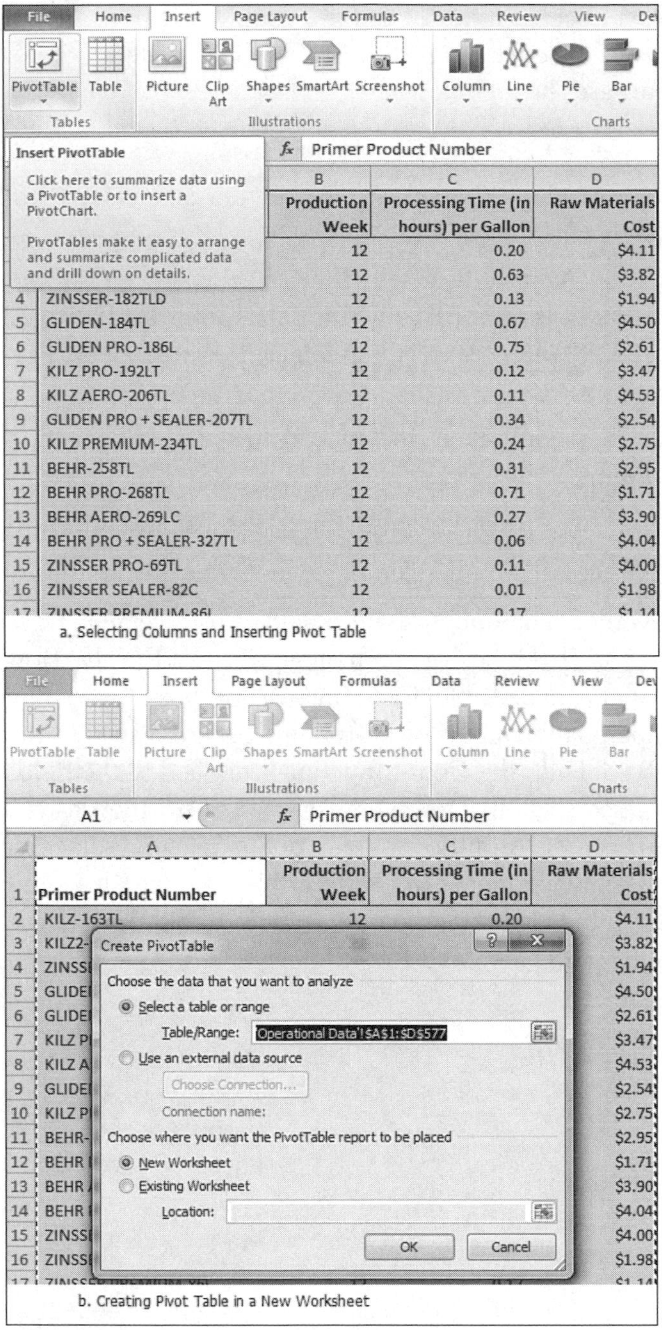

a. Selecting Columns and Inserting Pivot Table

b. Creating Pivot Table in a New Worksheet

Figure 3-2   Calculating the input parameters with a pivot table

	A	B	C	PivotTable Field List ▾ ✕
			Average of	Choose fields to add to report: 🔲 ▾
		Average of Raw	**Processing Time (in**	
1	**Primer Product Number** ▾	**Materials Cost**	**hours) per Gallon**	☑ **Primer Product Number**
2	BEHR AERO-269LC	$3.17	0.34	☐ Production Week
3	BEHR AERO-B41	$3.19	0.46	☑ **Processing Time (in hours) per Gall...**
4	BEHR AERO-B82	$4.17	0.42	☑ **Raw Materials Cost**
5	BEHR PRO + SEALER-327TL	$3.84	0.44	
6	BEHR PRO + SEALER-B42	$2.35	0.47	
7	BEHR PRO + SEALER-B86HD	$4.31	0.32	
8	BEHR PRO-268TL	$3.48	0.46	
9	BEHR PRO-B40	$4.34	0.39	
10	BEHR PRO-B81	$3.40	0.52	
11	BEHR-258TL	$4.28	0.31	
12	BEHR-B39	$3.74	0.51	
13	BEHR-B80N	$5.37	0.35	Drag fields between areas below:
14	GILDEN PREMIUM-86LTORG	$2.59	0.37	▼ Report Filter ⊞ Column Labels
15	GILDEN PREMIUM-B64	$2.84	0.42	Σ Values ▾
16	GLIDEN PRO + SEALER-207TL	$3.95	0.37	
17	GLIDEN PRO + SEALER-B29	$2.41	0.35	
18	GLIDEN PRO + SEALER-B76	$2.78	0.36	⊞ Row Labels Σ Values
19	GLIDEN PRO-186L	$3.18	0.53	Primer Produc... ▾ Average of R... ▾
20	GLIDEN PRO-B128	$4.20	0.39	Average of P... ▾
21	GLIDEN PRO-B70	$4.15	0.36	
22	GLIDEN-184TL	$4.76	0.46	
23	GLIDEN-B127	$4.12	0.40	☐ Defer Layout Update Update

Figure 3-3 The results of the pivot table

The results of the pivot table for average processing time and average cost of raw materials are then transferred to the next worksheet where the *contribution coefficients* are calculated for each primer, as shown in Figure 3-4. The contribution coefficients are calculated as a difference between price per gallon (column C) and production cost. Price per gallon is determined based on the supply and demand, historical data, and is also provided by the Sales Department based on the contractual agreement that PMI has with its vendors.

Production cost is equal to the average cost of raw materials plus the cost of labor. The cost of labor is calculated by multiplying the average processing time (column D) and the labor cost per hour, which in the case of PMI is $10. Contribution coefficients per gallon (*cc*) are located in the last column and are calculated as follows:

$$cc = price - (raw\ materials\ cost + processing\ times\ in\ hours \times labor\ cost\ per\ hour)$$

So for the first primer in the list (BEHR AERO-269LC), the contribution coefficient is calculated as F2 = C2 – (E2 + D2 × 10).

	A	B	C	D	E	F	G
			Price per	Processing	Raw Materials	Contribution	
1	Primer Product Number	j-index	Gallon	time (in hours)	Cost	Coefficients	
2	BEHR AERO-269LC	1	$10.20	0.34	$3.17	$3.64	<--=C2-(E2+D2*10)
3	BEHR AERO-B41	2	$8.75	0.46	$3.19	$0.97	
4	BEHR AERO-B82	3	$8.46	0.42	$4.17	$0.08	
5	BEHR PRO + SEALER-327TL	4	$8.26	0.44	$3.84	($0.02)	
6	BEHR PRO + SEALER-B42	5	$12.63	0.47	$2.35	$5.58	
7	BEHR PRO + SEALER-B86HD	6	$14.27	0.32	$4.31	$6.81	
8	BEHR PRO-268TL	7	$8.85	0.46	$3.48	$0.74	
9	BEHR PRO-B40	8	$6.10	0.39	$4.34	($2.12)	
10	BEHR PRO-B81	9	$9.01	0.52	$3.40	$0.43	
11	BEHR-258TL	10	$14.42	0.31	$4.28	$7.08	
12	BEHR-B39	11	$7.79	0.51	$3.74	($1.01)	
13	BEHR-B80N	12	$14.99	0.35	$5.37	$6.11	
14	GILDEN PREMIUM-86LTORG	13	$7.53	0.37	$2.59	$1.21	
15	GILDEN PREMIUM-B64	14	$10.42	0.42	$2.84	$3.37	
16	GLIDEN PRO + SEALER-207TL	15	$7.18	0.37	$3.95	($0.47)	
17	GLIDEN PRO + SEALER-B29	16	$12.22	0.35	$2.41	$6.36	
18	GLIDEN PRO + SEALER-B76	17	$9.99	0.36	$2.78	$3.64	
19	GLIDEN PRO-186L	18	$9.67	0.53	$3.18	$1.21	
20	GLIDEN PRO-B128	19	$7.25	0.39	$4.20	($0.84)	
21	GLIDEN PRO-B70	20	$12.94	0.36	$4.15	$5.21	
22	GLIDEN-184TL	21	$10.56	0.46	$4.76	$1.18	
23	GLIDEN-B127	22	$9.01	0.40	$4.12	$0.89	

Figure 3-4 Calculating contribution coefficients (partial view)

Step 2: Define Decision Variables

For large problems, decision variables are defined using general notations. For example, instead of stating that x_1 is the number of gallons of primer BEHR AERO-269LC to be produced in the production line during the week, or x_2 is the number of gallons of primer BEHR AERO-B41 163TL to be produced in the production line during the week, and so on, the modeler can generally state that:

$$x_j = \text{number of gallons of primer } j \text{ to be produced}$$
$$\text{in the production line during the week}$$

where j = 1, 2, 3, ..., 48

The LP parameters calculated via the pivot table are then transferred in a new worksheet. Note that this worksheet is named LP Modeling in the *ch3_PaintPrimer* file you can download from the companion website for this demonstration. Column C of this worksheet (as shown in Figure 3-5) holds the values of decision variables, initially set to one. Other operational data such as Maximum Production Level, Minimum Production Level, Available Machine Hours, and Available Raw Materials Budget are also placed in this worksheet.

Step 3: Formulate the Objective Function

The objective function is the sum of the product between the decision variables and the contribution coefficients. The SUMPRODUCT() function is used to calculate the total profit and is placed in cell H7 as follows:

$$H7 = SUMPRODUCT\ (\$C\$2:\$C\$49,\ D2:D49)$$

where $\$C\$2:\$C\$49$ is the range holding the decision variables and D2:D49 is the range holding contribution coefficients.

Step 4: Identify the Set of Constraints

There are several sets of constraints in the LP model for the PMI case. One set of constraints is related to the amount of hours the production lines are available during the week. PMI has a weekly capacity of 8,500 available machine hours. As a result, the first constraint of the LP model is:

$$0.34x_1 + 0.466x_2 + 0.42x_3 + ... + 0.42x_{48} \leq 8500$$

The left-hand side value of the equation is calculated in cell I7 (see Figure 3-5) using the following SUMPRODUCT() function:

$$I7 = SUMPRODUCT\ (\$C\$2:\$C\$49,\ E2:E49)$$

where $\$C\$2:\$C\$49$ is holding the values of the decision variables and E2:E49 is holding the values of the average processing time per gallon. The right-hand side value is stored in cell I2.

	A	B	C	D	E	F	G	H	I	J
1	Primer Product Number	j-index	Decision Variables	Contribution Coefficients	Processing time (in hours)	Raw Materials Cost	Maximum Production Level	Minimum Production Level	Available Machine Hours	Available Raw Materials Budget
2	BEHR AERO-269LC	1	1	$3.64	0.34	$3.17	1,200	200	8,500	$ 90,000
3	BEHR AERO-B41	2	1	$0.97	0.46	$3.19				
4	BEHR AERO-B82	3	1	$0.08	0.42	$4.17				
5	BEHR PRO + SEALER-327TL	4	1	($0.02)	0.44	$3.84				
6	BEHR PRO + SEALER-B42	5	1	$5.58	0.47	$2.35		Total Profit	Machine Hours	Raw Materials Purchases
7	BEHR PRO + SEALER-B86HD	6	1	$6.81	0.32	$4.31		$94.86	19.36	$172.44
8	BEHR PRO-268TL	7	1	$0.74	0.46	$3.48	=SUMPRODUCT($C$2:$C$49,D2:D49)			
9	BEHR PRO-B40	8	1	($2.12)	0.39	$4.34				
10	BEHR PRO-B81	9	1	$0.43	0.52	$3.40				
11	BEHR-258TL	10	1	$7.08	0.31	$4.28				
12	BEHR-B39	11	1	($1.01)	0.51	$3.74				
13	BEHR-B80N	12	1	$6.11	0.35	$5.37				
14	GILDEN PREMIUM-86LTORG	13	1	$1.21	0.37	$2.59				
15	GILDEN PREMIUM-B64	14	1	$3.37	0.42	$2.84				
16	GILDEN PRO + SEALER-207TL	15	1	($0.47)	0.37	$3.95				
17	GILDEN PRO + SEALER-B29	16	1	$6.36	0.35	$2.41				
18	GILDEN PRO + SEALER-B76	17	1	$3.64	0.36	$2.78				
19	GILDEN PRO-186L	18	1	$1.21	0.53	$3.18				
20	GILDEN PRO-B128	19	1	($0.84)	0.39	$4.20				
21	GILDEN PRO-B70	20	1	$5.21	0.36	$4.15				
22	GILDEN-184TL	21	1	$1.18	0.46	$4.76				
23	GILDEN-B127	22	1	$0.89	0.40	$4.12				

Figure 3-5 Template for the paint primer production mix model

The second set of constraints is related to the maximum production quantity for each type of paint primer. The requirement that PMI produces no more than 1,200 gallons per week for each product is established via a set of constraints as follows:

$$x_1 \le 1200, x_2 \le 1200, x_3 \le 1200, ..., x_{48} \le 1200$$

Similarly, the minimum production requirement of no less than 200 gallons for each product can be expressed:

$$x_1 \ge 200, x_2 \ge 200, x_3 \ge 200, ..., x_{48} \ge 200$$

The final set of constraints is related to the budget limitation for purchasing weekly raw materials. PMI has a weekly budget of $9,000 and this value is stored in cell J2. The constraint is:

$$3.17x_1 + 3.19x_2 + 4.17x_3 + \cdots + 5.19x_{48} \le 9000$$

The left-hand side value of the equation is calculated in cell J7 as follows:

$$J7 = SUMPRODUCT (\$C\$2:\$C\$49, F2:F49)$$

where $\$C\$2:\$C\$49$ is holding the values of the decision variables and F2:F49 is holding the values of the average raw materials cost per gallon.

Step 5: Identify the Set of Non-Negativity Constraints

Finally, the LP model must be forced to generate non-negativity values for the decision variables with the following constraints:

$$x_1 \ge 0, x_2 \ge 0, x_3 \ge 0 ... x_{48} \ge 0$$

Solving Linear Programming Models with Excel

After the Excel template is completed, Solver can be invoked to solve the PMI LP model. As in the case of problems with two decision variables, the same steps can also be followed to solve larger LP models with Solver:

1. Set up constraints and objective functions in Solver.

2. Generate the solution and results.

3. Use sensitivity analysis to gain greater insight.

Step 1: Set Up Constraints and Objective Function in Solver

After Solver is opened, the Solver Parameters dialog box shown in Figure 3-6 opens. In the Set Objective field, enter the objective cell **$H$7** (total profit). This is the value Solver will attempt to maximize. Next, select the Changing Våariable Cells field to enter the decision variables (**$C$2:$C$49**). Enter the constraints, explained previously, by clicking the Add button. For more information on how to add constraints, refer to Appendix B, "A Brief Tour of Solver."

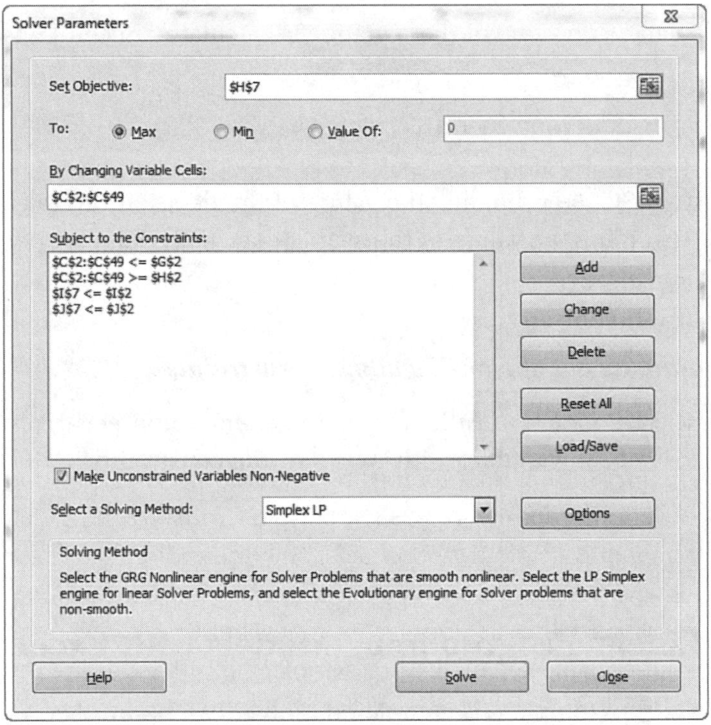

Figure 3-6   The Solver Parameters dialog box

Step 2: Generate the Solution and Results

Solver generates an Answer Report, which shows details of the solutions: the value of the objective function (cell E16 in Figure 3-7), the final values for the decision variables that determine the objective function (cells E21: E68 in Figure 3-7), and the state of each constraint for the given solution (shown in Figure 3-8). For example, the Answer Report (Figure 3-7) indicates that the PMI can potentially generate revenues equal to $92,358.64 per week. This level of revenue can be achieved when the company produces certain quantities of each type of primer. Specifically, the company must produce 915 gallons of primer BEHR AERO-269LC, 200 gallons of BEHR AERO-B41, 200 gallons of BEHR AERO-B82, 1,200 gallons of BEHR PRO + SEALER-B42, and so on.

A B		C	D	E	F
14 Objective Cell (Max)					
15	Cell	Name	Original Value	Final Value	
16	$H$7	BEHR PRO + SEALER-B86HD Total Profit	$94.86	$92,358.64	
17					
18					
19 Variable Cells					
20	Cell	Name	Original Value	Final Value	Integer
21	$C$2	BEHR AERO-269LC Decision Variables	1	915	Contin
22	$C$3	BEHR AERO-B41 Decision Variables	1	200	Contin
23	$C$4	BEHR AERO-B82 Decision Variables	1	200	Contin
24	$C$5	BEHR PRO + SEALER-327TL Decision Variables	1	200	Contin
25	$C$6	BEHR PRO + SEALER-B42 Decision Variables	1	1,200	Contin
26	$C$7	BEHR PRO + SEALER-B86HD Decision Variables	1	1,200	Contin
27	$C$8	BEHR PRO-268TL Decision Variables	1	200	Contin
28	$C$9	BEHR PRO-B40 Decision Variables	1	200	Contin
29	$C$10	BEHR PRO-B81 Decision Variables	1	200	Contin
30	$C$11	BEHR-258TL Decision Variables	1	1,200	Contin
31	$C$12	BEHR-B39 Decision Variables	1	200	Contin
32	$C$13	BEHR-B80N Decision Variables	1	1,200	Contin
33	$C$14	GILDEN PREMIUM-86LTORG Decision Variables	1	200	Contin
34	$C$15	GILDEN PREMIUM-B64 Decision Variables	1	200	Contin
35	$C$16	GLIDEN PRO + SEALER-207TL Decision Variables	1	200	Contin
36	$C$17	GLIDEN PRO + SEALER-B29 Decision Variables	1	1,200	Contin
37	$C$18	GLIDEN PRO + SEALER-B76 Decision Variables	1	200	Contin
38	$C$19	GLIDEN PRO-186L Decision Variables	1	200	Contin

Figure 3-7   Answer Report—objective function and decision variables

	Cell	Name	Cell Value	Formula	Status	Slack
71	Constraints					
72	Cell	Name	Cell Value	Formula	Status	Slack
73	$I$7	Machine Hours	8500	$I$7<=$I$2	Binding	0
74	$J$7	Raw Materials Purchases	79218.13	$J$7<=$J$2	Not Binding	10781.87295
75	$C$2	BEHR AERO-269LC Decision Variables	915	$C$2<=$G$2	Not Binding	285.4679803
76	$C$3	BEHR AERO-B41 Decision Variables	200	$C$3<=$G$2	Not Binding	1000
77	$C$4	BEHR AERO-B82 Decision Variables	200	$C$4<=$G$2	Not Binding	1000
78	$C$5	BEHR PRO + SEALER-327TL Decision Variables	200	$C$5<=$G$2	Not Binding	1000
79	$C$6	BEHR PRO + SEALER-B42 Decision Variables	1,200	$C$6<=$G$2	Binding	0
80	$C$7	BEHR PRO + SEALER-B86HD Decision Variables	1,200	$C$7<=$G$2	Binding	0
81	$C$8	BEHR PRO-268TL Decision Variables	200	$C$8<=$G$2	Not Binding	1000
82	$C$9	BEHR PRO-B40 Decision Variables	200	$C$9<=$G$2	Not Binding	1000
83	$C$10	BEHR PRO-B81 Decision Variables	200	$C$10<=$G$2	Not Binding	1000
84	$C$11	BEHR-258TL Decision Variables	1,200	$C$11<=$G$2	Binding	0
85	$C$12	BEHR-B39 Decision Variables	200	$C$12<=$G$2	Not Binding	1000
86	$C$13	BEHR-B80N Decision Variables	1,200	$C$13<=$G$2	Binding	0
87	$C$14	GILDEN PREMIUM-86LTORG Decision Variables	200	$C$14<=$G$2	Not Binding	1000
88	$C$15	GILDEN PREMIUM-B64 Decision Variables	200	$C$15<=$G$2	Not Binding	1000
89	$C$16	GLIDEN PRO + SEALER-207TL Decision Variables	200	$C$16<=$G$2	Not Binding	1000
90	$C$17	GLIDEN PRO + SEALER-B29 Decision Variables	1,200	$C$17<=$G$2	Binding	0
91	$C$18	GLIDEN PRO + SEALER-B76 Decision Variables	200	$C$18<=$G$2	Not Binding	1000
92	$C$19	GLIDEN PRO-186L Decision Variables	200	$C$19<=$G$2	Not Binding	1000
93	$C$20	GLIDEN PRO-B128 Decision Variables	200	$C$20<=$G$2	Not Binding	1000
94	$C$21	GLIDEN PRO-B70 Decision Variables	1,200	$C$21<=$G$2	Binding	0

Figure 3-8 Answer Report—constraints

The Answer Report (Figure 3-8) also indicates the status of each constraint. For example, Machine Hours is indicated as a *binding* constraint. That means that producing the suggested quantities for each primer will fully utilize the available 8,500 hours of machine time. On the contrary, Raw Materials Purchases is not a binding constraint. Out of $90,000 available to purchase raw materials, the optimal solution requires that PMI purchases only $79,218.13, which leaves a slack of unused $10,781.87.

Decision makers can use the analysis of constraints to make better decisions. Normally, a *not binding* constraint indicates a partial usage of the resource. Adding more resources in this constraint does not change the value of the objective function. That is a waste of resources if PMI decides to allocate more money to the Raw Materials Purchases budget because the existing budget is not fully utilized yet. On the other hand, a binding constraint indicates a full usage of the resource or a full satisfaction of the constraint. If the factory needs to make more profit, more machine hours must be added through buying new machines or extending the hours of the existing machines (adding another shift, for example). Also, because the company does not want to produce more than the weekly demand, binding constraints (see rows 79 and 80) also mean that the factory has produced the maximum possible allowance for these two primers.

Step 3: Use Sensitivity Analysis to Gain Greater Insight

Sensitivity analysis is an important tool in gaining additional insights about the model output. The sensitivity analysis allows the decision maker to identify the conditions under which the changes in the input contribution coefficients or the right-hand side values will not alter the solution. Given the importance of this analysis for generating business intelligence from the LP model output, the next two sections use the PMI case to demonstrate the effect of changes in the right-hand side values and changes in the contribution coefficients.

Sensitivity Analysis: Changes in the Right-Hand Side Values

This section shows how the decision maker can interpret the Solver Sensitivity Report for the PMI problem. Figure 3-9 illustrates the output of the Sensitivity Report for the constraints of this LP model. There are two constraints in the model: Machine Hours (binding) and Raw Materials Purchases (not binding).

58 Constraints						
59			Final	Shadow	Constraint	Allowable	Allowable
60	Cell	Name	Value	Price	R.H. Side	Increase	Decrease
61	$I$7	Machine Hours	8500	10.76402065	8500	96.58333333	241.75
62	$J$7	Raw Materials Purchases	79218.13	0	90000	1E+30	10781.87295

Figure 3-9 Sensitivity Report for the constraints

The *shadow price* indicates how the value of the objective function changes when one additional unit of the constraint is acquired. Suppose that instead of having 8,500 machine hours, the right-hand side value is increased to 8,501 machine hours. How would that affect the objective function? Instead of rerunning the solution, the decision maker can simply look at the shadow price and know that the objective function will increase by exactly $10.76. Alternatively, when the available machine hours are reduced by one hour, the shadow price indicates that the value of the objective function will be reduced by $10.76.

The allowable increase and allowable decrease values can be used to determine the range of right-hand side values where the shadow price impact remains true. This information is useful and allows the decision maker to create several would-be scenarios to find the value of the optimal solution without having to rerun the model. That is, the right-hand side value for the machine hours can increase up to 96.58 units to a total of 8,596.58 hours and the objective function (total profit) can improve by $10.76 per each unit. This means that the new profit can be calculated as long as the total machine hours range from the lower limit 8,258.25 (8,500 – 241.75) to the upper limit 8,596.58. So, for example, if the machine hours increase from 8,500 to 8,580 (two additional machines × 40 per week per machine), there will be a total profit increase of $860.80 (80 × $10.76) for an overall total profit of $92,358.64 + $860.80 ≈ $93,219. On the other side, if machine hours are reduced by 100 hours, the total profit will be reduced by $1,076 dollars.

Shadow prices have the same meaning for not binding constraints. In the case of the Raw Materials Purchases constraint, the shadow price is zero, the allowable increase is 1E+30 (infinity), and the allowable decrease is 10,781.87. That means, the right-hand side value of the purchasing budget can increase beyond $90,000 with no impact in the objective function. After all, the existing budget is not completely used in the optimal solution. On the other side, the budget can decrease down to approximately $79,218 and the solution will still remain the same. The decision maker needs to rerun the model when the right-hand side values of the constraints change beyond the lower and upper limits.

Decision variables also have upper bounds and lower bounds and can be seen as constraints. For example, in the case of the PMI model, the requirement was not to produce more than 1,200 gallons of each primer and to produce at least 200 gallons for each primer. Even when there is no simple lower bound, decision variables must always comply with the non-negativity constraint, so the lower bound is zero. The limits of the decision variables are shown in the Limits Report and indicated in Figure 3-10.

				Lower	Objective	Upper	Objective
6		Objective					
7	Cell	Name	Value				
8	$H$7	Total Profit	$92,358.64				
9							
10							
11		Variable		Lower	Objective	Upper	Objective
12	Cell	Name	Value	Limit	Result	Limit	Result
13	$C$2	BEHR AERO-269LC Decision Variables	915	200	89,756	915	92,359
14	$C$3	BEHR AERO-B41 Decision Variables	200	200	92,359	200	92,359
15	$C$4	BEHR AERO-B82 Decision Variables	200	200	92,359	200	92,359
16	$C$5	BEHR PRO + SEALER-327TL Decision Variables	200	200	92,359	200	92,359
17	$C$6	BEHR PRO + SEALER-B42 Decision Variables	1,200	200	86,778	1,200	92,359
18	$C$7	BEHR PRO + SEALER-B86HD Decision Variables	1,200	200	85,544	1,200	92,359
19	$C$8	BEHR PRO-268TL Decision Variables	200	200	92,359	200	92,359
20	$C$9	BEHR PRO-B40 Decision Variables	200	200	92,359	200	92,359
21	$C$10	BEHR PRO-B81 Decision Variables	200	200	92,359	200	92,359
22	$C$11	BEHR-258TL Decision Variables	1,200	200	85,282	1,200	92,359
23	$C$12	BEHR-B39 Decision Variables	200	200	92,359	200	92,359
24	$C$13	BEHR-B80N Decision Variables	1,200	200	86,244	1,200	92,359
25	$C$14	GILDEN PREMIUM-86LTORG Decision Variables	200	200	92,359	200	92,359
26	$C$15	GILDEN PREMIUM-B64 Decision Variables	200	200	92,359	200	92,359
27	$C$16	GLIDEN PRO + SEALER-207TL Decision Variables	200	200	92,359	200	92,359

Figure 3-10   Limits Report for the PMI model

Note the optimal value of the first decision variable 914.5 (rounded to 915). The constraint related to this decision variable can be considered as a not binding constraint because the right-hand side value of this constraint is the lower bound of 200. So, there is a slack of 715. If the value of the first decision variable (BEHR AERO-269LC) is reduced to 200, then the objective function will be reduced to the Objective Result, 89,756 (92,358 − 715 × 3.64) where 3.64 is the contribution coefficient of this variable. The optimal value of the second decision variable (BEHR AERO-B41) is 200. As such, the constraint related to the decision variable can be considered as a binding constraint. So there is no slack in this constraint—there is no room for this variable to change without impacting the objective function.

Sensitivity Analysis: Changes in the Contribution Coefficients

Solver also produces a Sensitivity Report for decision variables. Figure 3-11 shows a partial list of decision variables, their final value, reduced cost, objective coefficient, allowable increase, and allowable decrease. The final value indicates the final values of the optimal solution. The optimal solution indicates 914.5 (rounded to 915) gallons of primer BEHR AERO-269LC; 200 gallons of BEHR AERO-B41, BEHR AERO-B82, and BEHR PRO + SEALER-327TL; 1,200 gallons of BEHR PRO + SEALER-B42; and so on.

The reduced cost for a decision variable, by definition, is the shadow price of the constraint for that variable. As such, the reduced cost indicates how much the objective function changes if the constraint of the variable changes by one unit. The optimal solution indicates that the company must make 915 gallons of primer BEHR AERO-269LC. The reduced cost for this variable is 0, contribution coefficient is 3.641826988, allowable increase is 0.375508065, and allowable decrease is 0.091206755. That means that PMI will continue to make 915 gallons of this primer, with no impact in the value of the objective function (because the reduced cost is zero) as long as the contribution coefficient for this variable remains within the lower limit of 3.550620232 (3.641826988 − 0.091206755) and upper limit of 4.017335053 (3.641826988 + 0.375508065). The zero value for the reduced cost also indicates that the objective function will not change if the lower bound (right-hand side value) 200 of this variable changes.

Now, consider the second decision variable, BEHR AERO-B41. Because $x_2 \geq 200$ is a constraint in the LP formulation, then this constraint has no slack; it is a binding constraint and has a shadow price (in this case, a reduced cost) of −3.9613 (to round for four digits). Assume for a moment that the constraint is changed to $x_2 \geq 201$. In this case, the value in the objective function will be decreased by 3.9613 (increased by the reduced cost). Of course this is expected because the ideal solution was not to produce 200 units of this primer. Anything other than this amount will provide a less-than-optimal value for the objective function.

	Cell	Name	Final Value	Reduced Cost	Objective Coefficient	Allowable Increase	Allowable Decrease
7							
8							
9	$C$2	BEHR AERO-269LC Decision Variables	914.5320197	0	3.641826988	0.375508065	0.091206755
10	$C$3	BEHR AERO-B41 Decision Variables	200	-3.961359718	0.972149748	3.961359718	1E+30
11	$C$4	BEHR AERO-B82 Decision Variables	200	-4.456816186	0.077527515	4.456816186	1E+30
12	$C$5	BEHR PRO + SEALER-327TL Decision Variables	200	-4.79188979	-0.019840634	4.79188979	1E+30
13	$C$6	BEHR PRO + SEALER-B42 Decision Variables	1200	0.521641746	5.580731453	1E+30	0.521641746
14	$C$7	BEHR PRO + SEALER-B86HD Decision Variables	1200	3.423839446	6.814505952	1E+30	3.423839446
15	$C$8	BEHR PRO-268TL Decision Variables	200	-4.234185686	0.744173866	4.234185686	1E+30
16	$C$9	BEHR PRO-B40 Decision Variables	200	-6.287629182	-2.116571179	6.287629182	1E+30
17	$C$10	BEHR PRO-B81 Decision Variables	200	-5.154322127	0.425028579	5.154322127	1E+30
18	$C$11	BEHR-258TL Decision Variables	1200	3.775623714	7.076590047	1E+30	3.775623714
19	$C$12	BEHR-B39 Decision Variables	200	-6.457221537	-1.01242109	6.457221537	1E+30
20	$C$13	BEHR-B80N Decision Variables	1200	2.338322825	6.114700071	1E+30	2.338322825
21	$C$14	GILDEN PREMIUM-86LTORG Decision Variables	200	-2.808911413	1.209656297	2.808911413	1E+30
22	$C$15	GILDEN PREMIUM-B64 Decision Variables	200	-1.156812025	3.373046666	1.156812025	1E+30
23	$C$16	GILDEN PRO + SEALER-207TL Decision Variables	200	-4.450201762	-0.46751412	4.450201762	1E+30
24	$C$17	GILDEN PRO + SEALER-B29 Decision Variables	1200	2.648373076	6.361960202	1E+30	2.648373076
25	$C$18	GILDEN PRO + SEALER-B76 Decision Variables	200	-0.20901431	3.639123073	0.20901431	1E+30
26	$C$19	GILDEN PRO-186L Decision Variables	200	-4.4639646	1.214056294	4.4639646	1E+30
27	$C$20	GILDEN PRO-B128 Decision Variables	200	-5.031237287	-0.84223925	5.031237287	1E+30

Figure 3-11 Sensitivity Report for decision variables

Big Optimizations with Big Data

There are quintillions of bytes of data floating inside and outside today's organizations. The challenge is to turn this data into action and optimal decisions. "For decision-making factories today, the key raw material is information. The contemporary transformation, occurring in an environment of Big Data, is called big optimization."[32] Amazon and Google are good examples of using Big Data for big optimization for competitive advantage. Amazon continually tries to reduce the time between an order and its delivery.[33] Google's search engine optimization can capture, manipulate, and draw conclusions from large data sets and has paved the way for its *Big Data optimization* strategy.[32] However, small and medium-sized companies are also using the power of big optimization to gain a competitive advantage. A recent study [34] estimates that "one standard deviation increase in the use of business analytics correlates with an approximately five-to-six-percent improvement in productivity and a slightly higher percent improvement in profitability across a broad spectrum of companies."

The implementation of linear programming models as operational business intelligence tools in the era of Big Data requires that decision makers attract as many sources of data as possible, adapt that data from heterogeneous sources, and feed those models directly with the most up-to-date input parameters. Optimization models become an integral part of operational decision making and are driven by business processes.

Process-driven models require the implementation of *magnetism*, *agility*, and *depth*, known as the MAD approach.[35] A *magnetic* system attracts all sources of data, not just good data but also outliers, unstructured data, and missing values. This inclusive approach allows data scientists to incorporate all sources of data and all data points for a better representation of the state of the system. *Agility* implies the ability to produce and adapt data from heterogeneous sources. Dealing with heterogeneity of data sources, replacing missing values, and data imputation allows the decision scientist to deal with the variety of data. Relational database systems, declarative query languages, and specialized data marts with structured data can then be used to retrieve data input for LP models. A *deep* system supports complex statistical, machine learning, and optimization analysis. The depth approach to optimization models assumes a direct connection of the optimization packages with operational databases.

Finally, the implementation of LP models in the era of Big Data requires that data scientists consider a trade-off between less-than-optimal, but practical solutions and optimal, but complex and often delayed solutions. John Tukey of Princeton states that "an approximate answer to the right problem is worth a good deal more than an exact answer to an approximate problem."[32]

Wrap Up

The chapter suggests a step-by-step approach for formulating and solving large LP models. These steps can be organized into two major stages: formulating LP models and Solving LP models with Solver. The modeler can set up an Excel template that represents equations of the objective function and constraints. Then, the Solver's parameter box is invoked, allowing the program to generate results. This approach allows the modeler to connect the template with operational databases, capture and process a large amount of data, and continue to use these templates with the next round of data. In addition, due to the complexity of the large amount of input data into such problems, calculating parameters of the model is suggested in this chapter as an additional and first step. The pivot table from Excel is suggested as a good tool to summarize operational data into LP parameters and to always refresh those parameters with the latest operational data. The methodology offered in this chapter is graphically represented in Figure 3-12.

Formulating LP Models

- Step 1: Calculate model parameters with pivot table
- Step 2: Define decision variables
- Step 3: Formulate objective function
- Step 4: Identify the set of constraints
- Step 5: Identify the set of non-negativity constraints

Solving LP Models with Solver

- Step 1: Set up constraints and objective functions in Solver
- Step 2: Generate the solution and results
- Step 3: Use sensitivity analysis to gain greater insight
  - Analyze changes in the right-hand side values of the constraints
  - Analyze changes in the contribution coefficients

Figure 3-12   Two-step approach to solving large LP models

Review Questions

1. Discuss the advantages and the challenges of implementing linear programming models in real business settings. How can spreadsheet modeling be used to assist decision makers when processing operational data into LP model parameters?

2. Discuss the importance of data preparation in the process of formulating LP models in the era of Big Data. Mention several Excel-based techniques that can be used to clean data, replace missing data, summarize data, and refresh data.

3. Discuss the advantages of using a pivot table for data preparation and parameter calculations. Why is the pivot table's Refresh option important?

4. Discuss the importance of understanding the output of LP models and being able to analyze the Sensitivity Report generated by Solver.

5. What are contribution coefficients and technological coefficients, and what is their role in a typical LP formulation?

6. What is the difference between a binding and not binding constraint? Is there a relationship between a binding (or not binding) constraint and slack?

7. What is the impact of the changes in the right-hand values of the constraints? Discuss the shadow price concept and how it affects the value of the objective function.

8. What is the impact of the changes in the contribution coefficients? Discuss the reduced cost concept and how it affects the value of the objective function.

9. Explain the MAD approach and how it can be used to explore Big Data with LP models.

10. Would you rather reach an approximate, but practical answer to a simplified LP model, or would you rather seek an exact answer with a complicated model that is not practical to formulate and solve? Provide examples to support your position.

Practice Problems

1. A chair furniture manufacturer makes 21 different types of chairs, which are categorized into single-seat chairs or multiple-seat chairs. The manufacturer can make at most three times as many single-seat chairs as multiple-seat chairs. There are 5,000 machine hours and $15,000 production budget available each month. The manufacturer has recorded data for 12 months regarding its production quantities, raw materials costs, total machine hours used, and total sales for each week for each chair. This data is stored in the Excel file named *ch3_P1chairproduction* and can be downloaded from the companion website at www.informit.com/title/9780133760354.

   a. Use a pivot table to process the operational data and calculate:

   • The average cost of raw materials for a single chair and a multiple chair

   • The average machine time for a single chair and a multiple chair

- The average sales price for a single chair and a multiple chair

b. Determine how many units for each category (single-seat and multiple-seat chairs) the company should produce during the next month to maximize the total expected sales:

- Formulate an LP model to represent the above optimization problem.
- Create an Excel template to calculate the total sales, machine hours, and cost of raw materials for a given production mix.
- Apply Solver to generate an optimal solution.
- Generate an Answer Report, Sensitivity Report, and Limits Report and provide recommendations for possible alternative production scenarios.

2. Use the same Excel file (*ch3_P1chairproduction*) and the same problem description provided in the problem 1 to calculate LP parameters for *each* of the 21 chairs and determine how many units of each individual product should be produced during the next month. The company must produce at least 10 units of each chair to allow for production variety and to meet contractual agreements. Perform a similar analysis with an Excel template and Solver. Compare the results with those from problem 1.

3. You are the production manager at a golf club manufacturer. The company wants to determine how many iron sets to produce each month so the company can maximize the revenue. The company has allocated 20,000 machine hours to iron sets. Operational data from the last six months is recorded and can be found in the *ch_P3ironsetsproduction* file, which can be downloaded from the companion website. This data consists of average processing time, cost of raw materials, and monthly demand for the last six months. The file also contains the retail price for each set, which can be used to calculate the contribution coefficients for each set. To keep contractual agreements, the company must produce at least 500 units, but no more than 1,000 units for each iron set. The company employs workers in the iron set assembly line and each of them is paid an average of $40 per hour.

a. Using a pivot table, process the Excel file data to calculate the average processing time, cost of raw materials, and monthly demand for each iron set.

b. Prepare an Excel template that calculates the values of the average net profit for each iron set, average time usage per each iron set, the total profit for a given production set, and the actual usage of machine hours for a given production mix. Assume an initial production level of one unit for each set.

c. Use Solver to set up the objective function and constraints and generate an optimal solution.

d. Analyze the results using the Answer Report, Sensitivity Report, and Limits Report.

4. Use the same Excel file (*ch_P3ironsetsproduction*) and the same problem description provided in problem 3 to solve an optimization problem with a different objective function. Now, the operations manager seeks to minimize the total materials cost and must utilize the 20,000 hours of labor. The rest of the constraints remain the same as in problem 3.

Perform a similar analysis with an Excel template and Solver. Compare the results with those from problem 3. Which solution is better: the one that maximizes profit or the one that minimizes cost? Why would a manager be interested in fully utilizing the available labor hours?

5. You are the logistics manager of a distributor of frozen foods. The product line is composed of 40 distinct products and they have different production and shipping costs due to the varying sizes of boxes and locations in the warehouse. The company has employed 50 workers who manufacture the food products. They work 40 hours per week, regardless of demand. The company has set budgets for production, packing, and shipping at $4,000, $6,000, and $13,000, respectively. Four weeks of operational data are recorded and a summary is presented in the *ch3_P5frozenfood* file from the companion website. The file contains data about product price per unit, raw materials cost per unit, packing costs per unit, shipping costs per unit,

production costs per hour, production time per unit, and maximum possible quantity to be shipped as constrained by prior contractual agreements. The manager wants to determine how many units of each product should be produced, packed, and shipped to maximize the total net profit contributions.

a. Prepare an Excel template that calculates the input parameters for the LP model. Use the net profit for each product as the contribution coefficients (cc) for each product, which can be calculated as:

$$cc = \text{price per unit} - (\text{shipping cost per unit} + \text{packing cost per unit} + \text{production cost per hour} \times \text{production time per unit})$$

b. Use Solver to set up the objective function and constraints and generate an optimal solution. Analyze the results using the Answer Report, Sensitivity Report, and Limits Report and answer the following managerial questions:

- What is the maximum possible profit that the company can reach?

- How many units of each product must be produced, packed, and shipped for the optimal solution?

- If additional funding is made available, where should the company invest among its four resources: labor hours, production budget, packing budget, or shipping budget? Explain why.

6. Consider the same problem described in problem 5. As a production manager, you want to modify the model and you seek to minimize overall production, packing, and shipping costs. Also, the constraint related to the production labor hours is changed: The manager must utilize the available labor hours and avoid potential layoffs. Rerun the model and answer the following questions:

a. What is the minimum possible production, shipping, and packing cost?

b. How many units of each food must be prepared, packed, and shipped? Select two types of food that are not recommended to be produced by the final solution. How much should the contribution coefficients for these products change so they can become part of the optimal solution?

7. Colored Cosmetics is a mineral cosmetics company that ships their custom-blended products all over the world. The company produces various colors of lipsticks, eyeliners, eye shadows, blushes, and foundations. You are the operations manager for the company and your goal is to determine how many units of each product to produce each month to maximize revenue.

Operational data from the last three months is recorded and can be found in the *ch3_P7cosmetics* file, which can be downloaded from the companion website. This data consists of average processing time, cost of raw materials, and monthly demand for the last three months. There are currently 46 different color product combinations. Costs of producing each product vary depending on the different types of pigments used in the particular colors. There is a different cost for each product because certain pigments are more costly than others. A total of $3,000 is available every month to purchase raw materials. Also, a staff member puts in 10 hours per week (40 hours per month) at a labor cost of $12 per hour. The cosmetic facility wants to limit its production capacity to no more than the maximum demand for each line during the last three months.

a. Using a pivot table, process the Excel file data to calculate the average processing time, cost of raw materials, and maximum monthly demand for each cosmetic product.

b. Prepare an Excel template that calculates the values of the average net profit for each cosmetic product, average time usage per each cosmetic, total profit for a given product, and actual usage of labor hours for a given production mix. Assume an initial production level of one unit for each product.

c. Use Solver to set up the objective function and constraints and generate an optimal solution.

d. Analyze results using the Sensitivity Report and the Limits Report.

4

Business Analytics with Nonlinear Programming

Chapter Objectives

- Explain what nonlinear programming models are and why they are more difficult to solve than LP models
- Demonstrate how to formulate nonlinear programming models
- Show how to use Solver to reach solutions for nonlinear programming models
- Explain how to read the Answer Report and how to perform sensitivity analysis for nonlinear programming models
- Offer practical recommendations when using nonlinear models in the era of Big Data

Prescriptive Analytics in Action: Netherlands Increases Protection from Flooding[1]

Fifty-five percent of the Netherlands is susceptible to flood risk, and the government has spent almost one billion euro on building dikes and dunes. In 2008, the government received a recommendation from the Delta committee to increase protection standards tenfold. This suggestion was safe, but following it would be very expensive.

---

[1] https://www.informs.org/Sites/Getting-Started-With-Analytics/Analytics-Success-Stories/Case-Studies/Dutch-Delta-Commissioners

A second committee was asked to determine economically efficient flood protection standards for all dike ring areas. The goal was to estimate the optimal amount of protection standards for each dike in order to minimize the overall investment cost and ensure protection against high water and to maintain fresh water supplies for now and the future.

The committee formulated this problem as a nonlinear programming model and was able to find optimal standard levels for each of 53 dike ring areas in the country. The nonlinear programming model incorporated total long-term social costs, such as investment costs for heightening dikes and the expected loss of flooding considering dynamic effects of climate change and socioeconomic growth in the area. The solution suggested that it was not necessary to increase protection standards by a factor of ten. Many of the current protection standards were already acceptable. The solution suggested change to only three dike ring regions. The nonlinear programming model allowed the government to effectively identify the investment strategies and establish economically efficient flood protection standards with a much lower cost. It is estimated that the use of these models has saved about eight billion euros in investment costs.

Introduction

Accurate modeling of real-world business situations often involves nonlinear functions.[36] The linear programming models discussed in the previous chapter are based on the assumptions that the objective function and constraints are linear equations. By definition, any linear equation has to be both proportional and additive. However, business situations have relationships that are often not proportional or additive. For example, assume that the price per unit for product A is $4. If a company makes 1,000 units of this product, the gross revenue would be $4,000. If 100 units are made, the total revenue would be $400. Two thousand units would generate revenue equal to $8,000. However, this assumption may fail under the conditions of economies of scale. For example, if the company produces over 5,000 units, the price may drop to $3 per unit. The total revenue is no longer proportional to the production volume. Also, let's assume that the company makes 100 units of product A and 200 units of product B, which has a profit of $5 per unit. Normally, the total profit would be additive: $100 \times 4 + 200 \times 5$. Now assume that the company is running a "buy one, get the second item of equal or lower price for free." In that situation, the total profit is no longer additive and the total profit is simply 200×5 and the profit from product A is not added.

When the objective function or any of the constraints do not follow the proportionality or additivity requirement, the decision maker may choose to represent business relationships with a nonlinear programming (NLP) model. NLP models have the same structure as the LP models, that is, an objective function to be optimized and a set of constraints to be satisfied. However, as this chapter will later demonstrate, NLP models are more challenging to solve than the LP models. A decision maker may decide to ignore the assumptions of nonlinearity in exchange for a simpler and more practical solution. However, in many situations, the decision maker must choose NLP models because the choice of a more difficult, but more accurate solution obtained by an NLP model supersedes the choice of a simpler, but less representative solution of LP models.

A comprehensive view of nonlinear programming models, algorithms, and theories can be found in [36], [37], and [38]. Most recently, Griva et al. [39] discuss nonlinear programming models in the context of optimization techniques and in comparison to linear programming models. Also, the third edition of *Linear and Nonlinear Programming* [40] brings optimization models into the era of Big Data and offers modern theoretical insights.

Challenges to NLP Models

Linear relationships in the LP models can be represented by straight lines when there are two decision variables or planes when there are three decision variables. Nonlinear relationships in the NLP models can be represented with curved lines when the model has two decision variables or curved surfaces when the model has three decision variables. Visually, it is extremely difficult, even impossible, to represent relationships when LP or NLP models have a large number of decision variables—however, you can intuitively deduce that relationships in the NLP models are much more complicated to represent than those in the LP models.

Local Optimum Versus Global Optimum

The term *optimum* refers to either *maximum* or *minimum*, depending on the specific goal of the optimization model. For example, when you seek to optimize profit, the *optimum* value refers to the *maximum* profit. On the other hand, when you seek to optimize cost, the *optimum* value implies the *minimum* cost. Figure 4-1 assumes a maximization model and, as such, the terms optimum and maximum are used interchangeably. By definition, a *local optimum* is a point in the feasible region that has a better value than any other feasible point in the small neighborhood around it. To the contrary, a *global optimum* is the point with the best value in the area of feasible solutions. When the constraints are linear, any local optimum is indeed a global optimum as well. Recall that when solving LP models graphically, the objective function is moved parallel to itself until it reaches an extreme point of the area of feasible solutions.

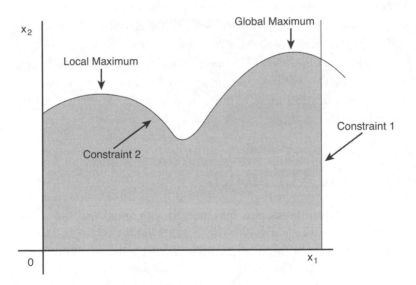

Figure 4-1 Area of feasible solutions with local and global maximum

That last point of contact is the point of optimal solution. That may not be the case for NLP models. Figure 4-1 shows the area of feasible solutions for an NLP model where one of the constraints, Constraint 2, is not linear. Once the objective function reaches the local maximum point, it is difficult to know whether that point is also a global maximum.

The Solution Is Not Always Found at an Extreme Point

As shown in the previous chapters, the solution of the LP models is found in the extreme points or border of the area of feasible solutions. The algorithms for LP models normally look for extreme points and compare the values of the objective function in those extreme points. That is not always the case for NLP models. Figure 4-2 shows the case of a maximization NLP model with linear constraints and a nonlinear objective function. As the objective function is gradually moved up (indicated by arrows), the last point of contact with the area of feasible solution is not in the corner point.

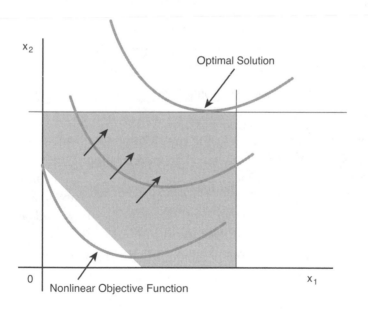

Figure 4-2 Possible optimal solution for NLP model

Multiple Feasible Areas

Consider the case shown in Figure 4-3 where there are three constraints. Nonlinear constraints may create discontinuous areas that satisfy all constraints. That makes solving the NLP models difficult because even if a solution algorithm is able to find the optimum within a particular feasible region, there is no guarantee that a better solution is not in some other feasible areas.

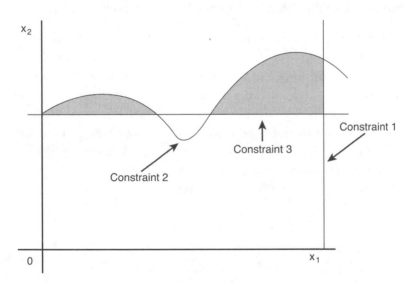

Figure 4-3 Multiple areas of feasible solutions

The above three situations describe graphically why NLP models are relatively more difficult to solve compared with LP models. Advanced heuristics, such as genetic algorithms, simulated annealing, generalized reduced gradient (GRG) method (available in Solver), quadratic programming, and barrier methods are developed to deal with such challenges. However, these algorithms add complexity to the solution process and often are not successful in finding or validating a global maximum.

Example 1: World Class Furniture

World Class Furniture (WCF) is a national retailer of home furniture with warehouses across the country. The WCF's warehouse stores five different furniture categories: tables, chairs, sofas, beds, and bookcases. The logistics manager wants to calculate the weekly order quantity for each furniture category to minimize overall inventory cost. The warehouse operates under the traditional Economic Order Quantity (EOQ) model. This model allows the manager to optimally calculate the amount of inventory for each item group with the goal of minimizing the total inventory cost. Inventory cost includes the storage (holding) cost and ordering cost. However, the EOQ model does not consider business constraints such as the storage capacity (200,000 cubic feet) and the purchasing budget ($1.5 million). Operational data about the inventory management for these five products is shown in Table 4-1.

Table 4-1 Operational Data for the Inventory Management System at WCF

Warehouse capacity (cubic feet)	200,000				
Average inventory budget	$1,500,000				

	Chairs	Tables	Beds	Sofas	Bookcases
Weekly demand (units)	1125	2750	3075	3075	750
Purchase price per unit	$45	$85	$125	$155	$125
Holding cost (per unit per week)	$2	$3	$3	$3	$4
Ordering cost (per order)	$100	$225	$135	$135	$100
Storage space required (cubic feet per unit)	84	106	140	70	100

Formulation of NLP Models

The same LP formulation steps can be used in the case of NLP models. Recall the following order of steps from Chapter 2, "Introduction to Linear Programming":

- Step 1: Define decision variables.
- Step 2: Formulate the objective function.
- Step 3: Identify the set of constraints.
- Step 4: Identify the set of non-negativity constraints.

Step 1: Define Decision Variables

The description of the WCF inventory management example states, *"The logistics manager wants to calculate the weekly order quantity for each furniture category in order to minimize overall inventory cost."* As such, the decision variables can be defined as:

- x_1 = number of tables to be ordered every week
- x_2 = number of chairs to be ordered every week
- x_3 = number of beds to be ordered every week
- x_4 = number of sofas to be ordered every week
- x_5 = number of bookcases to be ordered every week

Step 2: Formulate the Objective Function

The goal of the WCF manager is to *minimize* the inventory cost. Inventory cost is the sum of holding cost (H) plus ordering cost (O) per week. Holding cost can be calculated as the product between holding cost per unit (h) and the average level of holding inventory. Assuming a constant consumption for the products, the average holding inventory is half of the amount of inventory ordered. As a result, the holding cost for all products is:

$$H = \sum_{j=1}^{5} h_j \frac{x_j}{2} = 2\frac{x_1}{2} + 3\frac{x_2}{2} + 3\frac{x_3}{2} + 3\frac{x_4}{2} + 4\frac{x_5}{2}$$

The ordering cost per week is calculated as the product of ordering cost per order (o) and the number of orders. The number of orders on the other side is calculated as a division between weekly demand

(D) and the amount of units in each order. So, the ordering cost for all products is:

$$O = \sum_{j=1}^{5} o_j \frac{D_j}{x_j} = 100 \frac{1125}{x_1} + 225 \frac{2750}{x_2} + 135 \frac{3075}{x_3} + 135 \frac{3075}{x_4} + 100 \frac{750}{x_5}$$

The total weekly inventory management cost Z is calculated as a nonlinear function:

$$Z = H + O = \sum_{j=1}^{5} \left(h_j \frac{x_j}{2} + o_j \frac{D_j}{x_j} \right)$$

Step 3: Identify the Set of Constraints

There are two major constraints stated in the problem description: warehouse capacity and purchasing budget. The actual usage of warehouse capacity can be calculated as the product between amount of units in each order and the storage capacity per unit (s_j). This amount should not exceed the warehouse capacity (C). As a result:

$$\sum_{j=1}^{5} s_j x_j \leq C$$

or

$$84x_1 + 106x_2 + 140x_3 + 70x_4 + 100x_5 \leq 200000$$

The next constraint is related to the total purchasing limit (P). The amount actually spent on the purchasing, holding, and ordering inventory is calculated as the total holding and ordering cost (Z) plus purchasing cost, which in turn can be calculated as the product of purchasing price (p_j) and total amount ordered (D). So the constraint can be expressed as:

$$\sum_{j=1}^{5} \left(h_j \frac{x_j}{2} + o_j \frac{D}{x_j} + p_j D_j \right) \leq P$$

Step 4: Identify the Set of Non-Negativity Constraints

The following five constraints ensure that the model will not provide a negative value for the amount of units in each order for each product:

$$x_1 \geq 0, x_2 \geq 0, x_3 \geq 0, x_4 \geq 0, \text{ and } x_5 \geq 0$$

Considering all the previous four steps, the overall formulation for WCF's warehouse inventory management system is summarized as:

$$\text{Minimize } Z = \sum_{j=1}^{5} \left(h_j \frac{x_j}{2} + o_j \frac{D_j}{x_j} \right) \text{ (nonlinear objective function)}$$

subject to:

$$\sum_{j=1}^{5} s_j x_j \leq C \qquad \text{(linear constraint)}$$

$$\sum_{j=1}^{5} \left(h_j \frac{x_j}{2} + o_j \frac{D}{x_j} + p_j D_j \right) \leq P \qquad \text{(linear constraint)}$$

$$x_j \geq 0 \text{ for all } j=1, 2, 3, 4, 5 \qquad \text{(non-negativity constraint)}$$

Solving NLP Models with Solver

Similar to LP models, there are three steps to follow when solving NLP models:

- Step 1: Create an Excel template.
- Step 2: Apply Solver.
- Step 3: Interpret the solution results.

Step 1: Create an Excel Template

Figure 4-4 shows the what-if template for the WCF inventory problem. The economic order quantity (EOQ) calculates the optimal order quantity, the amount in each order that minimizes the overall inventory cost. The following classic EOQ formula [41] can be used in cells B15, C15, D15, E15, and F15:

$$EOQ = \sqrt{\frac{2oD}{h}}$$

If there are no storage or budget constraints, the amounts shown in cells B15, C15, D15, E15, and F15, respectively 335, 642, 526, 526, and 194, would be the optimal order amounts. However, due to operational constraints, these values may not be feasible. The inventory manager must seek the best possible feasible solutions and store these values in cells B16:F16. The initial inventory amount shown in Figure 4-4 corresponds to a trial solution of ordering 10 units per order for each product, as indicated in cells B16 through F16.

The template also calculates the *average inventory, average number of orders per week, total supply available, maximum cubic feet storage required, ordering cost per week, holding cost per week, inventory operating cost per week*, and *total inventory value* for each product. The specific formulas used to calculate each of these values for the last product (Bookcases in column F) are shown in the respective adjacent cells (G17:G24). Finally, the following measures are calculated: *actual capacity usage* (H20), *ordering cost per week* (H21), *holding cost per week* (H22), *value of the objective function* (H23), and *total inventory value* (H24), which includes purchasing, ordering, and holding costs. The formulas for these cells are shown in the adjacent cells (I20:I24).

Step 2: Apply Solver

Figure 4-5 shows the Solver Parameters dialog box. As shown earlier, cell H23 holds the overall cost. Choose this cell as the objective function and choose the Min option for this cell. Then, select cells B16 through F16 to hold the decision variables of the model by placing these cells in the By Changing Variable Cells field. Next, add the two constraints about the total capacity (H20<=B3) and purchasing budget (H24<=B4) in the "Subject to the Constraints" section. Finally, check the Make Unconstrained Variables Non-Negative check box. Because this is a nonlinear program, select the GRG Nonlinear option from the Select a Solving Method drop-down box.

	A	B	C	D	E	F	G	H	I
3	Warehouse capacity (cubic feet)	200,000							
4	Average inventory budget	$1,500,000							
5									
6		Chairs	Tables	Beds	Sofas	Bookcases			
7	Weekly demand (units)	1125	2750	3075	3075	750			
8	Purchase price per unit	$45	$85	$125	$155	$125			
9	Holding cost (per unit per week)	$2	$3	$3	$3	$4			
10	Ordering cost (per order)	$100	$225	$135	$135	$100			
11	Storage space required (cubic feet per unit)	84	106	140	70	100			
12									
13	Calculations and results								
14		Chairs	Tables	Beds	Sofas	Bookcases		Totals	
15	Economic order quantity (EOQ)	335	642	526	526	194	<----=SQRT(2*F10*F7/F9)		
16	Optimal order quantity (decision variables)	10	10	10	10	10			
17	Average inventory	5	5	5	5	5	<----=F16/2		
18	Average number of orders per week	112.50	275.00	307.50	307.50	75.00	<----=F7/F16		
19	Total supply available	1125	2750	3075	3075	750	<----=F16*F18		
20	Maximum cubic feet storage required	840	1060	1400	700	1000	<----=F16*F11	5000	<--=SUM(B20:F20)
21	Ordering cost per week	$11,250	$61,875	$41,513	$41,513	$7,500	<----=F10*F18	$163,650	<--=SUM(B21:F21)
22	Holding cost per week	$10	$15	$15	$15	$20	<----=F9*F17	$75	<--=SUM(B22:F22)
23	Inventory operating cost per week	$11,260	$61,890	$41,528	$41,528	$7,520	<----=F21+F22	$163,725	<--=SUM(B23:F23)
24	Total inventory cost (ordering+holding +purchasing)	$61,885	$295,640	$425,903	$518,153	$101,270	<----=F7*F8+F23	$1,402,850	<--=SUM(B24:F24)

Figure 4-4 Excel spreadsheet model for WCF inventory management problem

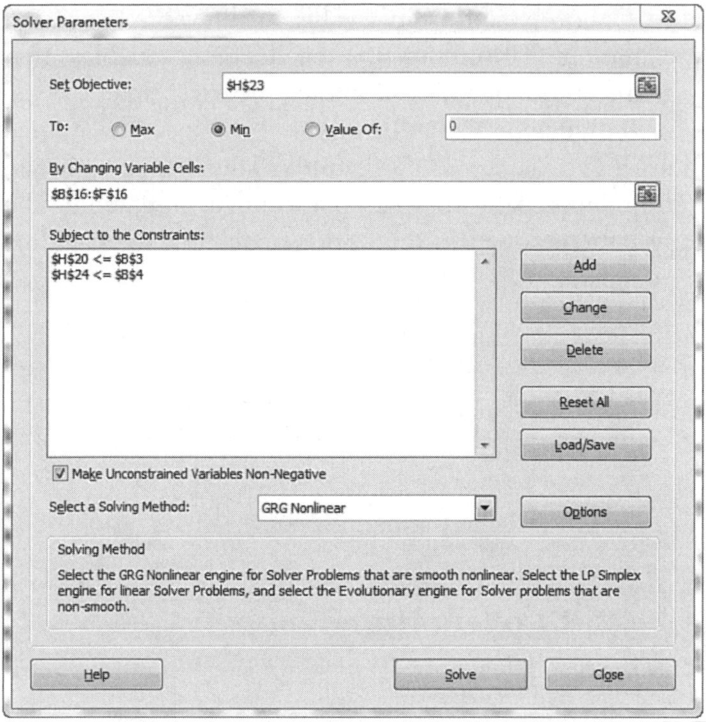

Figure 4-5 Solver Parameters dialog box for WCF inventory problem

Step 3: Interpret the Solution Results

After clicking the Solve button, the Solver Results dialog box appears. Select Keep Solver Solution and highlight the Answer and Sensitivity Reports. The Answer Report is shown in Figure 4-6. It contains three parts: Objective Cell, Variable Cells, and Constraints:

- **Objective Cell**—This cell indicates the final solution of the total inventory cost of $6,575. The original value, $163,725, is the value of the objective function for the initial trial solution when each furniture group is ordered ten units at a time.

- **Variable Cells**—Note the final value in the variable cells. It indicates the optimal solution for the decision variables, that is, the optimal order quantities for each product. Note that the values of the decision variables are smaller than the optimal quantities found via the EOQ formula. This indicates that in order to accommodate for space and budget limitations, the company has to order a less-than-optimal quantity for each furniture category.

A	B	C	D	E	F	G
14	Objective Cell (Min)					
15	Cell	Name	Original Value	Final Value		
16	$H$23		$163,725	$6,576		
17						
18						
19	Variable Cells					
20	Cell	Name	Original Value	Final Value	Integer	
21	$B$16	Optimal order quantity (decision variables) Chairs	10	289	Contin	
22	$C$16	Optimal order quantity (decision variables) Tables	10	575	Contin	
23	$D$16	Optimal order quantity (decision variables) Beds	10	457	Contin	
24	$E$16	Optimal order quantity (decision variables) Sofas	10	469	Contin	
25	$F$16	Optimal order quantity (decision variables) Bookcases	10	180	Contin	
26						
27						
28	Constraints					
29	Cell	Name	Cell Value	Formula	Status	Slack
30	$H$20		200000	$H$20<=$B$3	Not Binding	0.00011507
31	$H$24		$1,245,701	$H$24<=$B$4	Not Binding	254298.8745

Figure 4-6 Answer Report for WCF inventory problem

- **Constraints**—This section shows the left-hand side value for each constraint (under Cell Value), the direction of the constraint (under Formula), the Status, and the Slack. Figure 4-6 shows the warehouse storage capacity as a not binding constraint. However, the slack for this constraint, 0.000111507, is small and mainly due to the precision level selected in the Options section of Solver. It can easily be ignored or assumed to be zero. A constraint with a zero slack is a binding constraint. The total inventory and purchasing cost constraint, on the contrary, is truly not a binding constraint and has a slack of $254,298.

Sensitivity Analysis for NLP Models

There is a slight difference between the Sensitivity Reports of NLP models when compared with the Sensitivity Reports of the regular LP models. The *reduced cost* values in the case of LP models are called *reduced gradient* for NLP models. Recall that the reduced cost in the LP models is the shadow price for the decision variables. It shows how much the objective function will change if the right-hand side value of the respective constraint changes by one unit. As shown in Figure 4-7, the reduced gradients for the WCF problem are zero because the final value for each decision variables is positive.

6	Variable Cells			
7			Final	Reduced
8	Cell	Name	Value	Gradient
9	$B$16	Optimal order quantity (decision variables) Chairs	288.694089	0
10	$C$16	Optimal order quantity (decision variables) Tables	574.9289799	0
11	$D$16	Optimal order quantity (decision variables) Beds	457.2136075	0
12	$E$16	Optimal order quantity (decision variables) Sofas	469.2041824	0
13	$F$16	Optimal order quantity (decision variables) Bookcases	179.5302673	0
14				
15	Constraints			
16			Final	Lagrange
17	Cell	Name	Value	Multiplier
18	$H$20		199999.9999	-0.003470152
19	$H$24		$1,245,701.13	0

Figure 4-7 Sensitivity Report for the WCF inventory problem

The dual values for constraints are called shadow prices for LP models and Lagrange multipliers for NLP models. The Lagrange multiplier for the first constraint is very small, minus 0.003470152. This indicates how much the objective function will change if one more cubic foot is added in the warehouse capacity. Because the optimal solution uses the capacity fully, the objective function, the inventory management cost, can further be reduced by adding more capacity. The Lagrange multiplier for the second constraint is zero. Because this is not a binding constraint, increasing the budget further will have no effect on the value of the objective function.

A final note about sensitivity analysis for NLP models: The values of the reduced gradient and the Lagrange multiplier are valid only at the point of the optimal solution. The dual values change as soon as these values move away from the optimal solution. That is why there are no upper or lower limits (range of values) for NLP models. In the case of LP models, dual values remain constant within a range.

Example 2: Optimizing an Investment Portfolio

The trade-off between return on investment and risk is an important aspect in financial planning. Smart Investment Services (SIS) is an investment consulting firm that designs annuities, IRAs, 401(k) plans, and other products for investors with a variety of risk tolerances. The firm is currently preparing a portfolio involving a mix of eight mutual funds. SIS has compiled performance data for mutual funds during the last ten years. The rate of return on each mutual fund for each year is shown in Table 4-2. The Excel file for this exercise is named *ch4_portfolio* and can be downloaded from the companion website (www.informit.com/title/9780133760354). Assuming that the performance of these mutual funds over the past ten years is a good representative of the performance for the next 12 months, the firm wants to establish the portion of each mutual fund in the portfolio with the goal of maximizing the average return rate during the next year within a moderate level of risk.

Table 4-2 Mutual Funds' Performance During the Last Ten Years

Mutual Funds	Year 1	Year 2	Year 3	Year 4	Year 5	Year 6	Year 7	Year 8	Year 9	Year 10
Large-cap U.S. Growth	17%	9%	7%	-2%	-2%	8%	8%	16%	2%	8%
Mid-cap U.S. Growth	11%	13%	-2%	10%	4%	14%	4%	-4%	-9%	-4%
Small-cap U.S. Growth	8%	-13%	-3%	4%	18%	10%	13%	11%	10%	1%
Large-cap U.S. Value	13%	3%	-1%	7%	10%	13%	-6%	6%	8%	17%
Mid-cap U.S. Value	6%	-3%	-7%	18%	1%	5%	12%	12%	-5%	8%
Small-cap U.S. Value	13%	17%	12%	7%	14%	2%	2%	-2%	6%	2%
International Stock	6%	4%	9%	9%	3%	5%	5%	7%	10%	11%
Specialty Funds	8%	1%	6%	8%	8%	8%	-6%	-5%	6%	8%

Investment Portfolio Problem Formulation

The same steps will be followed in this example.

Step 1: Define Decision Variables

The description of the SIS investment decision states, *"The firm wants to establish the portion of each mutual fund in the portfolio with the goal of maximizing the average return rate during the next year within a moderate level of risk."* As such, the decision variables can be defined as:

- x_1 = proportion of portfolio invested in the large-cap U.S. growth during the next year
- x_2 = proportion of portfolio invested in the mid-cap U.S. growth during the next year
- x_3 = proportion of portfolio invested in small-cap U.S. growth during the next year
- x_4 = proportion of portfolio invested in large-cap U.S. value during the next year
- x_5 = proportion of portfolio invested in mid-cap U.S. value during the next year
- x_6 = proportion of portfolio invested in small-cap U.S. value during the next year
- x_7 = proportion of portfolio invested in international stock during the next year
- x_8 = proportion of portfolio invested in specialty funds during the next year

Step 2: Formulate the Objective Function

The firm wants to *maximize* the average rate of return. If the same portfolio were applied during the past ten years, then the rate of return in year i would be:

$$R_i = \sum_{j=1}^{8} r_{ij} x_j \text{ for all } i=1,2,...10$$

where:

- R_i is the rate of return in year i.
- r_{ij} is the rate of return of mutual fund j in year i.

For example, for the first two years, the rate of return is calculated as:

$R_1 = 17\%x_1 + 11\%x_2 + 8\%x_3 + 13\%x_4 + 6\%x_5 + 13\%x_6 + 6\%x_7 + 8\%x_8$

$R_2 = 9\%x_1 + 13\%x_2 - 13\%x_3 + 3\%x_4 - 3\%x_5 + 17\%x_6 + 4\%x_7 + 1\%x_8$

The average rate of return can be calculated as:

$$\bar{R} = \frac{R_1 + R_2 + ... + R_{10}}{10}$$

The objective function is then formulated as:

$$Maximize \ Z = \bar{R} = \frac{R_1 + R_2 + ... + R_{10}}{10}$$

Step 3: Identify the Set of Constraints

The first requirement is that the sum of the proportions must be equal to one. As such, the following constraint must be added to the formulation:

$$x_1 + x_2 + x_3 + x_4 + x_5 + x_6 + x_7 + x_8 = 1$$

The next requirement is that the portfolio must represent at least a moderate risk level. The investment risk is often associated with the variance of the portfolio's return. Because the ten-year scenarios are equally likely, the variance of the return can be calculated as:

$$Var = \frac{1}{10} \sum_{i=1}^{10} (R_i - \bar{R})^2$$

In other words, the variance of a given portfolio is the average of the sum of squares of the deviation of the rate of return of each year from the mean rate of return. Based on the historical data, a moderate level of risk is associated with a variance of 15. As such, the risk constraint can be stated as:

$$\frac{1}{10}\sum_{i=1}^{10}\left(R_i - \bar{R}\right)^2 \leq 15$$

Step 4: Identify the Set of Non-Negativity Constraints

The following eight constraints ensure that the model will not provide a negative value for the amount of units in each order for each product:

$$x_1 \geq 0, x_2 \geq 0, x_3 \geq 0, x_4 \geq 0, x_5 \geq 0, x_6 \geq 0, x_7 \geq 0, \text{ and } x_8 \geq 0$$

The overall formulation for the investment portfolio problem that seeks to maximize the return within a moderate risk level is summarized as:

$$Maximize \ \ Z = \frac{R_1 + R_2 + \ldots + R_{10}}{10} \qquad \text{(linear objective function)}$$

subject to:

$$x_1 + x_2 + x_3 + x_4 + x_5 + x_6 + x_7 + x_8 = 1 \qquad \text{(linear constraint)}$$

$$\frac{1}{10}\sum_{i=1}^{10}\left(R_i - \bar{R}\right)^2 \leq 15 \qquad \text{(nonlinear constraint)}$$

$$\sum_{j=1}^{8} r_{ij}x_j = R_i \text{ for all } i=1,2,\ldots 10 \qquad \text{(linear constraint)}$$

$$x_j \geq 0 \text{ for all } j=1, 2, \ldots, 10 \qquad \text{(non-negativity constraint)}$$

The objective of the NLP model is to maximize the average rate of return. Note that decision variables (x_j) are already included in the objective function because each return is a function of the set of decision variables. The first constraint ensures that all the money is invested in the mutual funds. The second constraint requires that investment risk is less than or equal to a moderate level. The third set of constraints defines the rate of return for each year. Finally, non-negativity constraints ensure that proportions of portfolio invested in each mutual fund are not negative numbers.

Solving the Portfolio Problem

Figure 4-8 shows the what-if template for this investment problem. The values for the proportion of the investment in each mutual fund (decision variables) are initially set to 12.5, an equal distribution between different options. The rate of return (shown in row 13 of the Excel template) is calculated as the sum of the product (SUMPRODUCT function) between the rates of return for funds for each year and their respective decision variables divided by eight. For example, the rate of return for Year 1 is the SUMPRODUCT (B3:B10, L3:L10)/8. Cell L13 holds the average of these rates and is selected as the Set Objective cell.

	A	B	C	D	E	F	G	H	I	J	K	L	
2	Mutal Funds	Year 1	Year 2	Year 3	Year 4	Year 5	Year 6	Year 7	Year 8	Year 9	Year 10	Decision Variables	
3	Large-cap U.S. Growth	17%	9%	7%	-2%	-2%	8%	8%	16%	2%	8%	12.50	
4	Mid-cap U.S. Growth	11%	13%	-2%	10%	4%	14%	4%	-4%	-9%	-4%	12.50	
5	Small-cap U.S. Growth	8%	-13%	-3%	4%	18%	10%	13%	11%	10%	1%	12.50	
6	Large-cap U.S. Value	13%	3%	-1%	7%	10%	13%	-6%	6%	8%	17%	12.50	
7	Mid-cap U.S. Value	6%	-3%	-7%	18%	1%	5%	12%	12%	-5%	8%	12.50	
8	Small-cap U.S. Value	13%	17%	12%	7%	14%	2%	2%	-2%	6%	2%	12.50	
9	International Stock	6%	4%	9%	9%	3%	5%	5%	7%	10%	11%	12.50	
10	Specialty Funds	8%	1%	6%	8%	8%	8%	-6%	-5%	6%	8%	12.50	
11												100	=SUM(L3:L10)
12												Average	
13	Rate of Return	10.25	3.88	2.63	7.63	7.00	8.13	4.00	5.13	3.50	6.38	5.85	=AVERAGE(B13:K13)
14	Variance	19.36	3.90	10.40	3.15	1.32	5.18	3.42	0.53	5.52	0.28	5.31	=AVERAGE(B14:K14)

=SUMPRODUCT(B3:B10,$L$3:$L$10)
=POWER((B13-$L$13),2)

Figure 4-8 Excel spreadsheet model for investment portfolio model

The variance is calculated in cell L14 as the average of cells B14 through K14, which in turn hold the difference between each rate of return with average rate of return in the second power. Cell L14 is limited to 15 and is shown as a constraint in the NLP model represented in the Solver Parameters dialog box shown in Figure 4-9. Also note that the non-negativity requirement for the decision variables is represented as a separate constraint in the Solver formulation, and, as a result, the Make Unconstrained Variables Non-Negative check box is left unchecked. It is suggested that you check the Use Multistart check box (found by clicking the Options button) to allow Solver to avoid local optimum as much as possible. Because this is a nonlinear program, select GRG Nonlinear from the Select a Solving Method drop-down list.

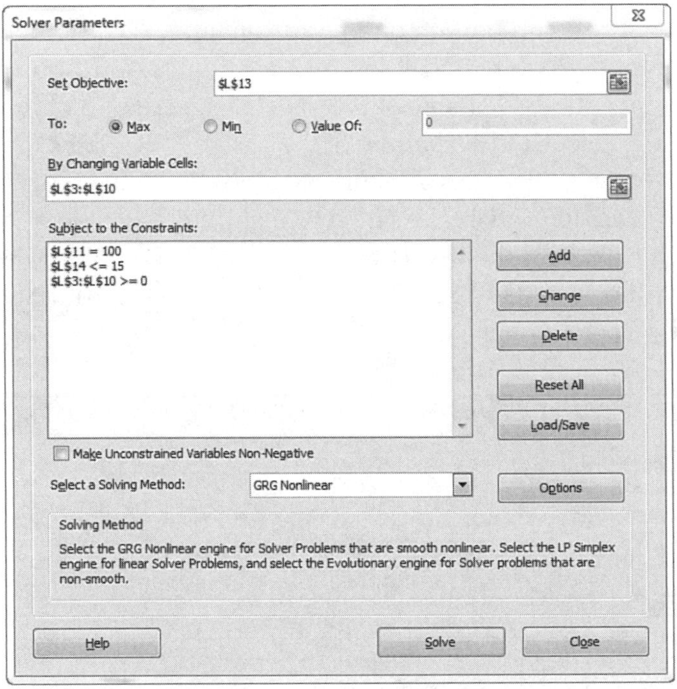

Figure 4-9 Solver Parameters dialog box for NLP SIS model

The Answer Report is shown in Figure 4-10. The final value in the Objective Cell section indicates a maximum rate of return, 7.22, suggested by Solver. The original value, 5.85, is the value of the objective function for the initial trial solution. This solution assumes an equal distribution of 12.5% for each mutual fund.

A	B	C	D	E	F	G
13						
14	Objective Cell (Max)					
15	Cell	Name	Original Value	Final Value		
16	$L$13	Rate of Return Average	5.85	7.22		
17						
18						
19	Variable Cells					
20	Cell	Name	Original Value	Final Value	Integer	
21	$L$3	Large-cap U.S. Growth Decision Variables	12.50	32.03	Contin	
22	$L$4	Mid-cap U.S. Growth Decision Variables	12.50	-	Contin	
23	$L$5	Small-cap U.S. Growth Decision Variables	12.50	-	Contin	
24	$L$6	Large-cap U.S. Value Decision Variables	12.50	5.85	Contin	
25	$L$7	Mid-cap U.S. Value Decision Variables	12.50	-	Contin	
26	$L$8	Small-cap U.S. Value Decision Variables	12.50	62.12	Contin	
27	$L$9	International Stock Decision Variables	12.50	-	Contin	
28	$L$10	Specialty Funds Decision Variables	12.50	-	Contin	
29						
30						
31	Constraints					
32	Cell	Name	Cell Value	Formula	Status	Slack
33	$L$11	Decision Variables	100	$L$11=100	Binding	0
34	$L$14	Variance Average	15.00	$L$14<=15	Binding	0
35	$L$3	Large-cap U.S. Growth Decision Variables	32.03	$L$3>=0	Not Binding	32.03
36	$L$4	Mid-cap U.S. Growth Decision Variables	-	$L$4>=0	Binding	-
37	$L$5	Small-cap U.S. Growth Decision Variables	-	$L$5>=0	Binding	-
38	$L$6	Large-cap U.S. Value Decision Variables	5.85	$L$6>=0	Not Binding	5.85
39	$L$7	Mid-cap U.S. Value Decision Variables	-	$L$7>=0	Binding	-
40	$L$8	Small-cap U.S. Value Decision Variables	62.12	$L$8>=0	Not Binding	62.12
41	$L$9	International Stock Decision Variables	-	$L$9>=0	Binding	-
42	$L$10	Specialty Funds Decision Variables	-	$L$10>=0	Binding	-

Figure 4-10 Answer Report for investment portfolio model

The final values in the Variable Cells section indicate the best investment strategy. Specifically, the maximum return can be achieved when the investor allocates 32.03% in large-cap U.S. growth, 5.85% in large-cap U.S. value, and 62.12% should be allocated in small-cap U.S. value mutual funds. The Constraints section indicates that the variance is a binding constraint and shows that the above optimal portfolio has a risk level of 15.

Exploring Big Data with Nonlinear Programming

The *volume* of Big Data offers increased opportunities for real-time intelligence and decision making. The availability of more data allows organizations to explore, formulate, and solve previously unsolvable problems.[42] However, the *variety* and *velocity* of Big Data offer significant challenges for optimization models. "Traditional linearization approaches, such as sampling a few values from the solution space, no longer deliver the business value required to stay competitive."[42] In the era of Big Data, advanced software programs, such as Solver, can be used to navigate trillions of permutations, variables, and constraints.

Wrap Up

This chapter discussed nonlinear optimization models. An NLP model has at least one nonlinear equation in either the constraint or the objective function. Many business applications can be represented by nonlinear functions. Nonlinear models can be used to capture the true nonlinear relationships between decisions and outcomes. The following list describes a few practical recommendations to follow when implementing NLP models:

- To simplify the formulation process, the same LP formulation steps can also be followed in the case of NLP models: define decision variables, formulate the objective function, identify the set of constraints, and identify the set of non-negativity constraints.

- Excel's Solver allows for easy switching among different algorithms. The GRG algorithm is best suited for NLP models. Also, selecting the Use Multistart option is suggested for greater efficiency when seeking the optimal solution and provides a better chance of finding a solution.

- There is always a risk that the algorithm will result in a local optimum. If the NLP problem can be reduced to a linear form, do so. Consider the trade-off between less-rigorous formulation and an efficient solution offered by LP versus a more accurate representation of nonlinear functions, but a less-efficient solution.

- Provide a good starting point in the trial template. The closer the initial starting point is to the eventual optimum point, the better chance there is that Solver will successfully find it. Use knowledge of the business domain to estimate the starting point. A decision maker may have already solved a similar problem using LP assumptions. Use that solution as you refine the objective function or constraints with new assumptions.

- Add a non-negativity constraint for decision variables instead of checking the non-negativity box. Searching algorithms often assigns zero values to decision variables. In the case of nonlinear functions, these zero values often create "division by zero" errors.

- Pay close attention when selecting Solver parameters. Read Appendix B, "A Brief Tour of Solver," for a detailed explanation of Solver parameters and their use in different scenarios.

A step-by-step methodology for using nonlinear programming models in practice is summarized in Figure 4-11.

Consider Potential Challenges of NLP Models

- Choose between LP models and NLP models.
- Nonlinear models are more difficult to solve.
- Nonlinear models offer better representation of nonlinear business relationships.

Formulate NLP Models

- Step 1: Define decision variables.
- Step 2: Formulate the objective function.
- Step 3: Identify the set of constraints.
- Step 4: Identify the set of non-negativity constraints.

Solve NLP Models with Solver

- Step 1: Create an Excel template.
- Step 2: Apply Solver.
- Step 3: Interpret the solution results.

Perform Sensitivity Analysis for NLP Models

- Reduced cost vs. reduced gradient
- Shadow price vs. Lagrange multiplier
- Upper and lower limits vs. valid only at point of optimal solution

Figure 4-11 Summary of NLP implementation methodology

Review Questions

1. Explain the assumptions of proportionality and additivity and how they may be violated in certain business situations. Provide examples to support your ideas.

2. What is the difference between a local optimum and a global optimum? Why might it be difficult to find a global optimum for NLP models?

3. Why are NLP models more difficult to solve than LP models? Graphically, what are some challenges that curved lines or surfaces present when trying to find an extreme point in the area of feasible solutions?

4. Perform an Internet search about one of the following methodologies: genetic search, simulated annealing, or GRG. Explain the algorithm used for seeking an optimal solution and why the selected approach is best fitted for NLP models.

5. What are the reduced gradient values in the sensitivity analysis of NLP models? How can a decision maker interpret the values of the Lagrange multipliers? What is the difference between reduced gradients and Lagrange multipliers, and how do they compare with the concept of shadow price from the LP models?

6. Why are the values of reduced gradients and Lagrange multipliers valid only at the point of optimal solution? Is there a range for them as in the case of LP models?

7. How can business applications be better represented using nonlinear optimization models? Should a decision maker ignore the nonlinearity assumption and search for a simpler solution methodology?

8. Describe the suggested steps followed to formulate an NLP. Are these different from LP formulation steps? Why or why not?

9. Can a data analyst know when a certain Solver solution is, in fact, a local or a global optimum? What can a data analyst do to avoid situations when a solution is trapped in the local optimum?

10. Why are solving Solver parameters important for the performance of the GRG algorithm? Why does selecting the Use Multistart check box help the quality of the final solution?

Practice Problems

1. A chair manufacturer makes two categories of chairs, single-seat chairs and multiple-seat chairs and wants to determine how many units for each category the company should produce during the next month. Each single-seat chair needs an average of $30 of raw materials and takes an average of 8 hours to make. Each multiple-seat chair needs an average of $70 of raw materials and takes an average of 10 hours to make. There are 10,000 hours available and a $20,000 budget to purchase raw materials each month. The fixed cost per unit decreases when the production volume increases. As a result, the profit for each single-seat chair and multiple-seat chair depends on the quantity of chairs made and is represented in the following table:

| *Single-Seat Chair* | | *Multiple-Seat Chair* | |
Order Quantity	**Profit per Unit**	**Order Quantity**	**Profit per Unit**
0–100	$40	0–200	$90
101–200	$50	201–300	$100
201–above	$70	301–above	$150

a. Formulate this problem as an NLP model. Use an IF statement to express the contribution coefficients as a function of quantity made. Explain why the model is nonlinear.

b. Create an Excel template to represent the relationships between decision variables, the objective function, and the set of constraints. Use the VLOOKUP function to express the net profit as a function of production quantities.

c. Use Solver to generate the solution and save the reports generated.

d. Interpret the Answer Report and perform a sensitivity analysis.

2. Using the same problem description as in problem 1, modify the model to include the additional requirement that the manufacturer must make at most three times as many single-seat chairs as multiple-seat chairs. Also, the company has started another production line and wants to allocate only 5,000 machine hours to the production of single-seat and multiple-seat chairs. Will the addition of new constraints change the solution? Does the reduction in machine hours impact the production quantity?

3. A paint manufacturer produces many primers, as shown in the *ch4_P3paintprimer* file, and wants to determine how many of each product should be made daily to maximize net profits. Due to contractual agreements, there is a maximum production level of 1,000 units per each primer, and a minimum production level of 500 units. There are 10,000 machine hours available and $100,000 budget for raw materials.

   The net profit for each component is a function of the amount of each primer to be produced. Specifically, the net profit can be calculated as:

$$NP = \left(\left(x_1^3 + x_2^3 + \ldots + x_n^3 \right) - 3\left(x_1^2 + x_2^2 + \ldots + x_n^2 \right) + 2\left(x_1 + x_2 + \ldots + x_n \right) \right) / 10,000$$

   where:

   - NP is the net profit for all products.
   - $x_1, x_2, \ldots x_n$ are the amounts of first, second, ... n^{th} primer to be produced per month.

   a. Formulate an NLP model that represents the above business description.

   b. Use Solver and generate an Answer Report and a Sensitivity Report.

   c. Perform a scenario analysis using the findings in the above reports.

4. Reconsider the toy manufacturing company that makes two toy trucks and toy cars, as described in Chapter 2. The manufacturing requirements for each toy production lot remain the same and are shown in the following table:

Raw Materials	Truck Toy	Car Toy	Available
Plastic	6 lb.	8 lb.	72 lb.
Labor hours	10 hrs.	8 hrs.	80 hrs.
Machine time	10 hrs.	4 hrs.	60 hrs.

The cost of producing T lots of toy trucks can be calculated as $700T + 40T^2 + 1,000$. The cost of producing C lots of toy cars is $200C + 20C^2 + 1,500$. There is a total budget of \$5,000 per week. The profit for either toy is \$500 per lot. The operational data can be found in the Excel file named *ch4_P4toys* and can be downloaded from the companion website (www.informit. com/title/9780133760354).

Formulate an NLP model that represents the above business description.

a. Solve the problems and indicate:

- What is the optimal number of toy truck and toy car lots to be purchased each month?
- What is the value of the objective function (total profit) for the above solution?
- Identify binding and not binding constraints for the optimal solution.

b. Perform a sensitivity analysis using Solver's output.

5. An appliance warehouse stocks the following items: micro-waves, ranges, washers, dryers, and dishwashers. Operational data in the *ch4_P5appliances* file, which can be downloaded from the companion website, indicates the demand for each appliance during the last 12 months. The following table stores additional operational data for each appliance:

	Microwave	Range	Washer	Dryer	Dishwasher
Monthly demand (units)	700	500	600	600	500
Selling price per unit	$300	$2,000	$2,000	$2,000	$800
Holding cost (per unit, period)	$15	$100	$100	$100	$40
Ordering cost (per order)	$70	$175	$200	$200	$185
Storage space required (cubic feet per unit)	3	18	20	22	18

You are the logistics manager and would like to calculate the monthly order quantity for each item category to minimize overall inventory cost. The Economic Order Quantity model can be used to optimally calculate the amount of inventory for each item group with the goal of minimizing the storage (holding) cost and ordering cost. However, there are some constraints such as storage capacity (5,000 cubic feet) and purchasing budget ($2.0 million) that may make the optimal solution of the EOQ model not feasible.

a. Formulate and solve the above problem as a linear programming model.

- What is the optimal order inventory level for each appliance? Are these quantities different from the values calculated via the EOQ formula? Why?

- What is the value of the objective function (holding plus ordering cost) for the above solution?

b. The company will implement a quantity discount policy when pricing each appliance. The following table represents the price for each appliance when the warehouse orders a specified quantity range. For example, if the warehouse orders up to 40 microwaves, the price is $160, but when the warehouse orders from 41 to 60 microwaves, the price drops to $150, and so on.

Adjust the Excel template to reflect the price discount using VLOOKUP functions.

Quantity Ordered	0–40	41–60	61–80	80 or over
Microwave	$160	$150	$140	$135
Range	$1,100	$1,000	$900	$950
Washer	$1,100	$1,000	$900	$950
Dryer	$1,100	$1,000	$900	$950
Dishwasher	$450	$400	$350	$300

   c. Formulate and solve the problem with an NLP model to determine the optimal order quantity that maximizes the total profit. Note that the profit for each appliance can be calculated as:

$$((\text{selling price} - \text{discounted purchasing price}) \times (\text{monthly demand})) - \text{total cost per month}$$

   d. Analyze the Answer Report to identify the final values for the decision variables, binding and not binding constraints, and the impact of changing the right-hand side values of the constraints in the final value of the objective function.

6. Consider the golf club manufacturer problem from the previous chapter. The goal of the operations manager is to determine how many iron sets to produce each month so the company can maximize the revenue. The amount of machine hours is still 20,000 and the company still pays its workers in the iron set assembly line an average of $40 per hour. Operational data for this exercise can be found in the *ch4_P6Iironsetsproduction* file, which can be downloaded from the companion website. The minimum production quota of 500 units and the maximum quota of 1,000 units are still required; however, the manager is allowed to offer a 20% discount to those iron sets that sell over 600 units per month.

   a. Explain why the new requirement will transform the model from LP to NLP.

    b. Recalculate the new contribution coefficients using an IF function.

    c. Calculate the values of the average net profit for each iron set, average time usage per each iron set, total profit for a given production set, and actual usage of machine and labor hours for a given production mix. Assume an initial production level of one unit for each set.

    d. Use Solver to set up the objective function and constraints and generate an optimal solution.

    e. Analyze results using the Sensitivity Report and the Limits Report.

7. Consider the same club manufacturing facility discussed in problem 6. In addition to suggested changes mentioned in problem 6, the operations manager has also concluded that the average machine hours per set is reduced by 0.1 hours (six minutes) when the production quantity increases over 300 units per set. This reduction is due to some elimination of the required setup time.

    The goal is to still identify the number of sets to be produced to maximize the profit.

    a. Adjust the Excel template to accommodate this requirement and explain why the adjusted constraint is nonlinear.

    b. Use Solver to set up the objective function and constraints and generate an optimal solution.

    c. Analyze results using the Sensitivity Report and the Limits Report.

    d. Solve the same problem but change the goal: Instead of maximizing the profit, the manager now seeks to minimize the production cost. Production cost in this case should include the cost of raw materials and labor cost. Interpret the result.

5

Business Analytics with Goal Programming

Chapter Objectives

- Discuss the importance of using goal programming models in business applications
- Demonstrate the process of formulating linear and nonlinear goal programming models
- Demonstrate the use of Solver for solving linear and nonlinear goal programming models
- Discuss the concept of aspiration levels and goal priorities
- Distinguish between functional variables and deviational variables in goal programming models
- Distinguish between systems constraints and goal programming constraints
- Offer practical recommendations for implementing goal programming models in business settings

Prescriptive Analytics in Action: Airbus Uses Multi-Objective Optimization Models[1]

Airbus is the world's leading aircraft manufacturer. Its jetliners range in size from low-cost and full-service carriers to the airfreight, VIP transport, and military airlift segments. Improving the product design and reducing product development time and cost are important goals for the company. These goals must be achieved under several constraints imposed by competition, regulatory environments, fuel efficiency, and customer expectations. As indicated in the company's website,[2] if the air traffic management system and technology on board the aircraft were optimized, "Every flight in the world could on average be around 13 minutes shorter... and ... this would save around 9 million tons of fuel annually which equates to over 28 million tons of CO_2 emissions." In 2013, Airbus undertook a multidisciplinary optimization modeling approach using optimization software known as MACROS. The software offers a user-friendly process integration, design optimization, and data analysis application. The program is used by data analysts, optimization experts, and engineers to perform trade-off studies more efficiently across all development phases. Some recent applications of MACROS at Airbus demonstrate the software's capability to "process nine objective functions, about a dozen degrees of freedom and more than thirty constraints in a timeframe compatible with conceptual design cycles." This tool has enabled engineers to find better design choices for the Airbus' family of aircraft with optimum performance relative to their respective seat and range capabilities.

---

[1] Airbus Press Release, "Airbus Achieves Multi-Objective Optimization of Its Aircraft Families with DATADVANCE's 'MACROS' software," August 28, 2013. [Online]. Available: http://www.airbus.com/newsevents/news-events-single/detail/airbus-achieves-multi-objective-optimization-of-its-aircraft-families-with-datadvances-macros/. [Accessed December 29, 2013.]

[2] http://www.airbus.com/innovation/future-by-airbus/

Introduction

Mathematical programming models discussed in the previous chapters seek to minimize or maximize a single objective function. This chapter discusses goal programming (GP), a mathematical programming methodology where the decision maker seeks to achieve not one, but multiple goals. Consider a mathematical programming model with a set of linear or nonlinear constraints. Now, assume that the decision maker in this problem seeks to achieve many goals, as in the case of Airbus where decision makers sought to achieve nine goals simultaneously.

The concept of GP was first introduced by Charnes, Cooper, and Ferguson in 1955.[43] The solution algorithm of GP and its first use as a decision analysis tool is described in the seminal works by Lee in 1972.[44] Lee's work was later expanded by Ignizio in 1974 [45] and Romero in 1991.[46] In 1995, Schniederjans [47] offered an up-to-date overview of GP models, the relationship between GP models to other management science techniques, practical recommendations for GP formulations and solutions, as well as a comprehensive bibliography of goal programming articles. More recently, Jones and Tamiz [48] provided an annotated bibliography of goal programming applications and wrote a textbook [49] in 2010 with a specific focus on the practical applications of GP models.

GP models have the potential to emerge as a primary data analytics tool for practitioners in the era of Big Data because goal programming models include multiple goals and are relatively easy to formulate and solve. For example, a linear GP model can be formulated and solved as a single LP model or in some cases as a series of connected LP models. This chapter demonstrates the solution methodology using a single model. Later in the book, Chapter 9, "Marketing Analytics with Multiple Goals," demonstrates the use of GP models in marketing campaigns and the solution approach proposed there combines several connected LP models. The value of the objective function in one model becomes a new constraint in the subsequent model until all optimization goals are incorporated.

GP Formulation

The process of formulating GP models is illustrated with the example of Rolls Bakery from Chapter 2, "Introduction to Linear Programming." Overall, the GP formulation has the following components:

- A minimization objective function
- A set of goal programming constraints
- An optional set of system constraints
- Non-negativity constraints for functional variables and deviational variables

Example 1: Rolls Bakery Revisited

Recall the details of the problem. The decision maker wants to determine how many dinner roll cases (DRC) and sandwich roll cases (SRC) to produce to maximize the net profit. The bakery has a total of 150 machine hours and each product is produced in lots of 1,000 cases. Products have a different wholesale price, processing time, cost of raw materials, and weekly market demand, as reproduced here in Table 5-1.

Recall from Chapter 2 that the LP formulation of this problem is:

$$\text{Max } Z = 400x_1 + 300x_2$$

subject to:

- $10x_1 + 15x_2 \leq 150$ (machine hours)
- $x_1 \geq 3$ (demand for product 1)
- $x_2 \geq 4$ (demand for product 2)
- $x_1, x_2 \geq 0$ (non-negativity constraints)

As shown in Chapter 2, the optimal solution for this problem suggests that the company must run nine lots of DRC and four lots of SRC. This production schedule will completely utilize the available 150 machine hours and generate the maximum possible net profit of $4,800.

Table 5-1 Production Requirements for One Lot of the Two Products

Product	(1) Wholesale Price per Case	(2) = (1) × 1,000 Wholesale Price per Lot	(3) Processing Time (in Hours) per Lot	(4) Cost of Raw Materials per Lot	(5) = (2)-(3) × 10-(4) Net Profit per Lot	(6) Demand for Cases	(7) = (6)/1000 Demand for Production Lots
DRC	$0.75	$750	10	$250	$400	3,000	3
SRC	$0.65	$650	15	$200	$300	4,000	4

The production manager at Rolls Bakery is revisiting the same problem with a new set of goals. Instead of simply maximizing the net profit, the manager wants to make sure that the following goals are met in the following order of priority:

- *Priority 1:* Company should not produce more than two lots over the weekly demand for each product (Goal 1).

- *Priority 2:* Company should meet the weekly demand for both products (Goal 2).

- *Priority 3:* Company should utilize available machine hours (Goal 3).

- *Priority 4:* Company should make the maximum possible net profit (Goal 4).

A few definitions will be helpful to the formulation of the GP model:

- **Aspiration level**—This is a specific value that indicates the desired or acceptable level of the objective. For example, the aspiration level for Goal 1 is five lots for DRCs and six lots for SRCs. The aspiration level for Goal 2 is three lots of DRCs and four lots of SRCs. The aspiration level for Goal 3 is 150 machine hours. The aspiration level for Goal 4 can be set to $4,800 because the optimal solution of the regular LP model led to $4,800 in profits.

- **Goal deviation**—This is the difference between the aspiration level and the actual accomplishment for each goal, as suggested by the GP solution. A goal is met when the goal deviation is zero. Ideally, all the deviations should be zero, but that is often not the case because the GP has several and often conflicting constraints. Because some of the goals cannot be achieved, the decision maker will try to achieve most of them in the order of their priority.

- **Goal priority**—Goal priority indicates the order of importance for achieving each goal. For example, the decision maker may decide that achieving the first and second goal is equally important. These two goals are much more important (15 times more) than the third goal, and the third goal is twice as important as the fourth goal. In that situation, the values for each priority are as follows:

- *Priority 1:* P1 = 300
- *Priority 2:* P2 = 300
- *Priority 3:* P3 = 20
- *Priority 4:* P4 = 10

Priority values are normally defined by the decision maker using his or her own experience or preferences in specific business settings. Sometimes, the priority values reflect potential penalties for not achieving the goal. For example, the penalty for not achieving the net profit by $1 is obviously $1. The penalty for overutilizing 150 machine hours is $2 per each additional hour in the form of overtime pay. Finally, assume that the penalty for not meeting or exceeding the demand for each unit is extremely important. Unsold products can be sold to a third party at an average loss of $30 per unit. In the GP formulation, the ratio between priorities is important, not the absolute differences between them. In other words, the set of priorities P1 = 300, P2 = 300, P3 = 20, and P4 = 10 is the same as the set P1 = 30, P2 = 30, P3 = 2, and P4 = 1 because the ratio is P1/P2 = 1, P1/P3 = 15, P2/P3 = 15, and P3/P4 = 2 for both sets of priorities.

GP Formulation Steps

The approach for reformulating an LP model into a GP model that incorporates the prioritized goals is relatively simple. The GP model requires a new set of decision variables. These variables represent underachievement or overachievement of a given goal. For example, because there is a priority to meet the market demand for each type of roll, and because these requirements are represented via Constraints 2 and 3 in the original LP model, then s_2^+ can be defined as the number of lots from DRC to be overproduced, and s_2^- can be defined as the number of lots from DRC to be underproduced. Similarly, s_3^+ can be defined as the number of lots from SRC to be overproduced, and s_3^- can be defined as the number of lots from SRC to be underproduced.

New constraints can be added to represent goals that are not currently represented by the existing constraints in the LP model. For example, the profit (objective function of the LP model) can be transformed into a constraint (Constraint 4) and the new goal programming decision variables could be defined as: s_4^+ is the amount of overachievement in the net profit and s_4^- is the amount of underachievement in the net profit.

s_i^+ and s_i^- are generally known as deviational variables. The decision maker needs to incorporate the deviational variables into a GP objective function and into the newly created or modified constraints. For example, when trying to minimize s_2^- and s_3^-, the decision maker is, in fact, seeking to avoid underproduction of DRC lots and SRC lots, which is actually the first goal with the highest priority in the GP problem. Similarly, when minimizing s_1^+ (the deviational variable for the machine hours constraints), the decision maker is seeking to avoid overusage of available hours, which is another goal of the second priority of the GP model.

This intuitive approach can be implemented using a more formal, step-by-step methodology for GP formulation. The suggested steps are:

- **Step 1:** Formulate the problem as a simple LP model.
- **Step 2:** Define deviational variables for each goal.
- **Step 3:** Write GP and system constraints.
- **Step 4:** Add non-negativity constraints for functional and deviational variables.
- **Step 5:** Determine the variables to be minimized in the objective function.
- **Step 6:** Write the objective function with priorities.

Step 1: Formulate the Problem as a Simple LP Model

The problem is first formulated as an LP model using the suggested methodology covered in the previous chapters. The decision variables in the LP model are known as functional variables and the constraints are known as functional or system constraints. Considering problem requirements, the Rolls Bakery problem is formulated as a maximization LP model, as shown in the previous chapter and reproduced at the beginning of this chapter.

However, considering Goal 1, the LP model is extended to include two more constraints:

$$\text{Max } Z = 400x_1 + 300x_2$$

subject to:

- $10x_1 + 15x_2 \le 150$ (machine hours)   (1)
- $x_1 \ge 3$ (demand for product 1)   (2)
- $x_2 \ge 4$ (demand for product 2)   (3)
- $x_1 \le 5$ (newly added constraint)   (4)
- $x_2 \le 6$ (newly added constraint)   (5)
- $x_1, x_2 \ge 0$ (non-negativity constraints)   (6)

Step 2: Define Deviational Variables for Each Goal

The decision maker is seeking to achieve four goals. The deviational variables for each goal can be defined as follows:

Goal 1: Company should not produce more than two lots over the weekly demand for each product.

Because this requirement is represented in Constraints (4) and (5), the following deviational variables are introduced:

- s_4^+ can be defined as the number of lots from DRCs to be overproduced when the aspiration level is five lots.
- s_4^- can be defined as the number of lots from DRCs to be underproduced when the aspiration level is five lots.
- s_5^+ can be defined as the number of lots from SRCs to be overproduced when the aspiration level is six lots.
- s_5^- can be defined as the number of lots from SRCs to be underproduced when the aspiration level is six lots.

Goal 2: Company should meet the weekly demand for both products.

This requirement is represented in Constraints (2) and (3). The following deviational variables are thus introduced:

- s_2^+ can be defined as the number of lots from DRCs to be overproduced when the aspiration level is three lots.

- s_2^- can be defined as the number of lots from DRCs to be underproduced when the aspiration level is three lots.

- s_3^+ can be defined as the number of lots from SRCs to be overproduced when the aspiration level is four lots.

- s_3^- can be defined as the number of lots from SRCs to be underproduced when the aspiration level is four lots.

Goal 3: Company should utilize available machine hours.

This requirement is represented in Constraint (1). The following deviational variables are introduced:

- s_1^+ can be defined as the number of machine hours over-utilized when the aspiration level is 150 hours.

- s_1^- can be defined as the number of machine hours underutilized when the aspiration level is 150 hours.

Goal 4: Company should make the maximum possible net profit.

This requirement is currently presented in the objective function of the LP. To be included in the GP model, the net profit must be transformed into a constraint. For now, assume this to be Constraint (6). The following deviational variables are introduced:

- s_6^+ can be defined as the amount of net profit to be overachieved when the aspiration level is $4,800.

- s_6^- can be defined as the amount of net profit to be underachieved when the aspiration level is $4,800.

Step 3: Write GP and System Constraints

System constraints are simply copied from the original LP. They do not reflect any specific goal, so no changes are made to those constraints. In the example under consideration, there are no system constraints; that is, all constraints in the LP incorporate one or more goals. GP constraints are written by incorporating deviational variables into the original LP constraints and using an equality sign as a direction. Specifically, the following GP constraints are created:

- $10x_1 + 15x_2 + s_1^- - s_1^+ = 150$ (machine hours)   (1)
- $x_1 + s_2^- - s_2^+ = 3$ (minimum demand for DRC)   (2)
- $x_2 + s_3^- - s_3^+ = 4$ (minimum demand for SRC)   (3)
- $x_1 + s_4^- - s_4^+ = 5$ (maximum demand for DRC) (4)
- $x_2 + s_5^- - s_5^+ = 6$ (maximum demand for SRC)   (5)
- $400x_1 + 300x_2 + s_6^- - s_6^+ = 4800$ (net profit constraint)   (6)

Step 4: Add Non-Negativity Constraints for Functional and Deviational Variables

Similar to LP models, the decision maker must enforce that all variables are zero or positive:

$$x_1, x_2, s_1^-, s_1^+, s_2^-, s_2^+, s_3^-, s_3^+, s_4^-, s_4^+, s_5^-, s_5^+, s_6^-, s_6^+ \geq 0$$

Step 5: Determine the Variables to Be Minimized in the Objective Function

Consider Constraint (1) from Step 3:

$$10x_1 + 15x_2 + s_1^- - s_1^+ = 150 \text{ (machine hours)}$$

When seeking to minimize the positive deviational variable s_1^+ the decision maker, in fact, is ensuring that the actual number of machine hours utilized is less 150 hours because negative deviation must be added to $10x_1 + 15x_2$ to reach 150. That means that the number of machine hours utilized will not exceed 150 hours. This is indeed what the decision maker wants to realize via Goal 3. As such, s_1^+ should be included as a variable to be minimized in the objective function. Following the same reasoning, this is the list of deviational variables to be included in the objective function of the GP model:

- Goal 1: s_4^+, s_5^+
- Goal 2: s_2^-, s_3^-
- Goal 3: s_1^+
- Goal 4: s_6^-

Step 6: Write the Objective Function with Priorities

Both Goal 1 and Goal 2 have the highest priority for the decision maker (P1 = P2 = 300). Goal 3 has the second-highest priority (P3 = 20) and Goal 4 has the third-highest priority (P 4= 10). As such, the objective function of the GP model can be written as:

$$Minimize\ Z = 300s_4^+ + 300s_5^+ + 300s_2^- + 300s_3^- + 20s_1^+ + 10s_6^-$$

Because the optimization algorithm will seek to minimize the value of Z, the first deviational variables to be reduced or even become zero are those that are associated with the largest values of contribution coefficients, those deviational variables that are associated with the highest priorities.

Putting It Together: GP Formulation for Rolls Bakery

After following these six steps, the completed GP formulation for the Rolls Bakery problems is presented as:

$$Minimize\ Z = 300s_4^+ + 300s_5^+ + 300s_2^- + 300s_3^- + 20s_1^+ + 10s_6^-$$

subject to:

- $10x_1 + 15x_2 + s_1^- - s_1^+ = 150$ (machine hours)   (1)
- $x_1 + s_2^- - s_2^+ = 3$ (minimum demand for DRC)   (2)
- $x_2 + s_3^- - s_3^+ = 4$ (minimum demand for SRC)   (3)
- $x_1 + s_4^- - s_4^+ = 5$ (maximum demand for DRC)   (4)
- $x_2 + s_5^- - s_5^+ = 6$ (maximum demand for SRC)   (5)
- $400x_1 + 300x_2 + s_6^- - s_6^+ = 4800$ (net profit constraint)   (6)

and

- $x_1, x_2, s_1^-, s_1^+, s_2^-, s_2^+, s_3^-, s_3^-, s_4^-, s_4^+, s_5^-, s_5^+, s_6^-, s_6^+ \geq 0$

Solving GP Models with Solver

As shown, the previous GP model is actually a minimization LP model with 14 decision variables and six constraints. Solution methodology is similar to the one followed for the regular LP models, as discussed in the previous chapters. Figure 5-1 indicates the setup of the Rolls Bakery GP model and its final solution achieved via Solver. The right section of the figure indicates the underlying formulas for each cell.

The values in the shaded cells indicate both functional variables (cells C8 and C9) and deviation variables (cells C12 through C23). Cells A12 through A23 indicate whether a priority is assigned or not to the deviational variables in cells B12 through B23. Priority values in cells D12 through D23 are calculated with a VLOOKUP function, which automatically assigns the priority values indicated in the area G12:H16. For example, the value in cell D12 is automatically calculated with VLOOKUP (A12, $G$12:$H$16, 2, FALSE). The aspiration levels are indicated in cells G6:G9 and I6:I7. These cells hold the *right-hand side* values of the GP constraints. The achievement levels in the adjacent cells (F6:F9 and H6:H7) represent the *left-hand side* values of the same constraints and are calculated as a combination of original LP constraints and their respective negative or positive deviational variables.

For example, the achievement level for the machine hours is placed in cell F8 and is calculated as E8+SUMPRODUCT(C12:C13, E12:E13). In this case, cell E8 is the actual usage of machine hours (165) and the SUMPRODUCT formula calculated the sum of product of deviational variables and their respective priority coefficients ("+1" for addition and "−1" for subtraction). In this case, the SUMPRODUCT value is −15. The objective function, placed in cell C24, is calculated as the sum of the product between deviational variables and their respective weight (priorities) using the following formula: SUMPRODUCT(C12:C23,D12:D23).

Top spreadsheet (values):

	Products	Wholesale Price per Case	Wholesale Price per Lot	Processing Time (in hours) per Lot	Cost of Raw Materials per Lot	Net Profit per Lot	Demand for Cases	Demand for Production Lots
	Dinner Roll Case (DRC)	$0.75	$750	10	$250	$400	3000	3
	Sandwich Roll Case (SRC)	$0.65	$650	15	$200	$300	4000	4

	Functional Variables	Number of Lots per Week	Net Profit Per Lot	Total Machine Time Used (in hours)	Achievement Level for Minimum	Aspiration Level for Minimum	Achievement Level for Maximum	Aspiration Level for Maximum
	DRC Lots (x1)	7.5	$3,000	75	3	3	5	5
	SRC Lots (x2)	6	$1,800	90	4	4	6	6
	Machine Hours			165	150	150	150	
	Profit		$4,800		4800	4800	4800	

Priority	Deviational Variables	Values	Priority Value				Priorities	
No Priority	s1negative	0	0	1			P1	300
P3	s1positive	15	20	-1			P2	300
P2	s2negative	0	300	1			P3	20
No Priority	s2positive	4.5	0	-1			P4	10
P2	s3negative	0	300	1			No Priority	0
No Priority	s3positive	2	0	-1				
No Priority	s4negative	0	0	1				
P1	s4positive	2.5	300	-1				
No Priority	s5negative	0	0	1				
P1	s5positive	0	300	1				
P4	s6negative	0	10	1				
No Priority	s6positive	0	0	-1				
	Objective Function		1050					

Bottom spreadsheet (formulas):

	Products	Wholesale Price per	Wholesale Price per Lot	Processing Time (in hours)	Cost of Raw Materials per Lot	Net Profit per Lot	Demand for Cases	Demand for Production
	Dinner Roll Case (DRC)	0.75	=C2*1000	10	250	=D2-E2*10-F2	3000	=H2/1000
	Sandwich Roll Case (SRC)	0.65	=C3*1000	15	200	=D3-E3*10-F3	4000	=H3/1000

	Functional Variables	Number of Lots per	Net Profit Per Lot	Total Machine Time Used (in	Achievement Level for Minimum	Aspiration Level for Minimum	Achievement Level for Maximum	Aspiration Level for
	DRC Lots (x1)	7.5	=C6*G2	=C6*E2	=C6+SUMPRODUCT(C14:C15,E14:E15)	=I2	=C6+SUMPRODUCT(C18:C19,E18:E19)	5
	SRC Lots (x2)	6	=C7*G3	=C7*E3	=C7+SUMPRODUCT(C16:C17,E16:E17)	=I3	=C7+SUMPRODUCT(C20:C21,E20:E21)	6
	Machine Hours			=E6+E7	=E8+SUMPRODUCT(C12:C13,E12:E13)	150	=I8+SUMPRODUCT(C22:C23,E22:E23)	
	Profit		=D6+D7		=D9+SUMPRODUCT(C22:C23,E22:E23)	4800		

Priority	Deviational Variables	Values	Priority Value				Priorities	
No Priority	s1negative	0	=VLOOKUP(A12,$G$12:$H$16,2,FALSE)	1			P1	300
P3	s1positive	15	=VLOOKUP(A13,$G$12:$H$16,2,FALSE)	-1			P2	300
P2	s2negative	0	=VLOOKUP(A14,$G$12:$H$16,2,FALSE)	1			P3	20
No Priority	s2positive	4.5	=VLOOKUP(A15,$G$12:$H$16,2,FALSE)	-1			P4	10
P2	s3negative	0	=VLOOKUP(A16,$G$12:$H$16,2,FALSE)	1			No Priority	0
No Priority	s3positive	2	=VLOOKUP(A17,$G$12:$H$16,2,FALSE)	-1				
No Priority	s4negative	0	=VLOOKUP(A18,$G$12:$H$16,2,FALSE)	1				
P1	s4positive	2.5	=VLOOKUP(A19,$G$12:$H$16,2,FALSE)	-1				
No Priority	s5negative	0	=VLOOKUP(A20,$G$12:$H$16,2,FALSE)	1				
P1	s5positive	0	=VLOOKUP(A21,$G$12:$H$16,2,FALSE)	1				
P4	s6negative	0	=VLOOKUP(A22,$G$12:$H$16,2,FALSE)	1				
No Priority	s6positive	0	=VLOOKUP(A23,$G$12:$H$16,2,FALSE)	-1				
	Objective Function	=SUMPRODUCT						

Figure 5-1 Model setup and solution for the GP model

Figure 5-2 shows the Solver setup for the same problem. Note the selection of Min for the objective function and the selection of functional and deviational variables in the By Changing Variable Cells field. Also, all GP constraints are set as equality constraints and the Make Unconstrained Variables Non-Negative check box is checked. The simplex LP is used as a solution method because this is a linear GP model.

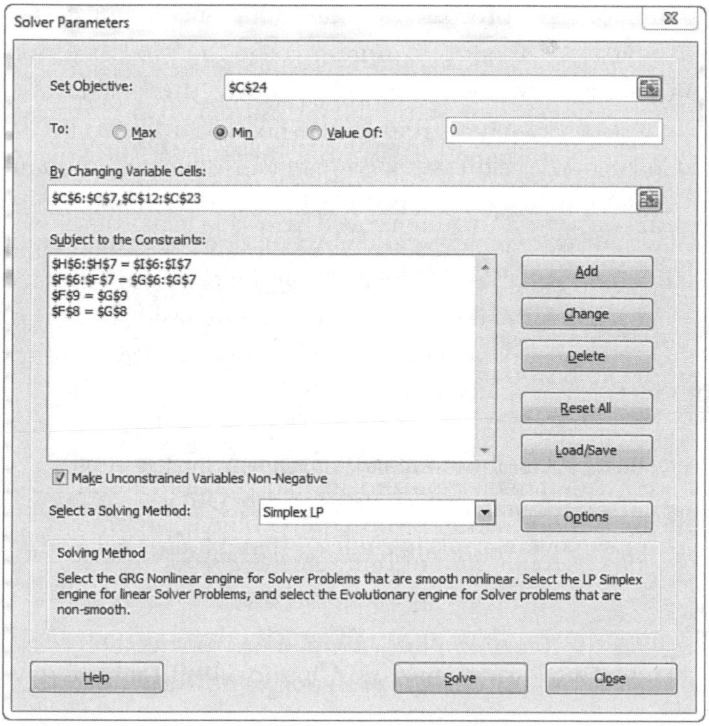

Figure 5-2 Solver setup for the GP model

The final solution indicates that the bakery must produce 7.5 lots of DRCs and 6 lots of SRCs. The values of the deviational variables indicate whether the decision maker has reached the stated goals. For example, recall that s_1^+, noted as s1positive in Figure 5-1 (under the Deviational Variables column), indicates the overutilization of machine hours. Because the value of s_1^+ resulted in 15, that shows that the optimal production of rolls required an additional 15 hours to produce for a total of $150 + 15 = 165$. Similarly, s4positive (s_4^+) is 2.5

and shows that the goal of not exceeding five production lots for DRC cases was not achieved. For a closer look at the results, the *ch5_bakery* file can be downloaded from the companion website.

Example 2: World Class Furniture

Nonlinear programming models can also be transformed into nonlinear GP models. The process is similar to the one demonstrated for the linear GP case. This section demonstrates the process of formulation and solutions for nonlinear GP models. Reconsider the inventory management example from World Class Furniture (WCF) discussed in Chapter 4, "Business Analytics with Nonlinear Programming." The logistics manager wants to calculate the weekly order quantity for each furniture category. The warehouse operates under the traditional Economic Order Quantity (EOQ) model and is limited by its storage capacity of 200,000 cubic feet and purchasing budget of $1.5 million. Operational data about the inventory management for these five products is reproduced in Table 5-2.

Table 5-2 Operational Data for the Inventory Management System at WCF

Warehouse capacity (cubic feet)	200,000				
Average inventory budget	$1,500,000				

	Chairs	Tables	Beds	Sofas	Bookcases
Weekly demand (units)	1125	2750	3075	3075	750
Purchase price per unit	$45	$85	$125	$155	$125
Holding cost (per unit per week)	$2	$3	$3	$3	$4
Ordering cost (per order)	$100	$225	$135	$135	$100
Storage space required (cubic feet per unit)	84	106	140	70	100

Recall the NLP formulation of the problem as follows:

$$Z = \sum_{j=1}^{5} \left(h_j \frac{x_j}{2} + o_j \frac{D_j}{x_j} \right)$$

(nonlinear objective function seeking to minimize the overall inventory holding and ordering cost)

subject to:

$$\sum_{j=1}^{5} s_j x_j \leq C$$

(linear constraint limiting the storage use to the maximum warehouse capacity C)

$$\sum_{j=1}^{5} \left(h_j \frac{x_j}{2} + o_j \frac{D}{x_j} + p_j D_j \right) \leq P$$

(nonlinear constraint limiting the total storage and purchasing cost to budget P)

$x_j \geq 0$ for all $j = 1, 2, 3, 4, 5$ (non-negativity constraint)

The optimal solution for this problem suggested that the warehouse must order 289 tables, 575 chairs, 457 beds, 469 sofas, and 180 bookcases. This solution reduced the total inventory costs to $6,576. The solution suggested that the warehouse storage capacity is a binding constraint, and that total inventory and purchasing cost constraints are not binding constraints and have a slack of $254,299.

Priorities and GP Formulation

After reaching this optimal solution, the logistics manager wants to explore more alternatives. For example, the company is interested in ordering four chairs for every table, and as such maintaining a 1 to 4 ratio between the number of tables and the number of chairs in each order. This requirement has a very high priority (Priority 1) and will allow the warehouse to meet future orders for sets that combine tables and chairs. In addition, the logistics manager would prefer (Priority 2) to not fully use the capacity of the warehouse and allow for unexpected orders to be accommodated on an ad hoc basis.

Finally, the company is still interested (Priority 3) in minimizing the overall cost of ordering and holding costs. However, to better achieve the first two goals, the company is willing to relax the cost minimization requirement and pay up to $7,000 instead of the minimum cost of $6,576, as suggested by the optimal solution of the original NLP model. To summarize, the logistics manager established the following three goals:

- *Goal 1:* Maintain a one to four ratio between tables and chairs (P1 = 1,000).
- *Goal 2:* Avoid overutilization of warehouse capacity (P2 = 50).
- *Goal 3:* Avoid spending more than $7,000 in holding and ordering costs (P3 = 1).

Deviational variables can be defined as follows:

- s_1^+ = the degree that the ratio between tables and chairs is over the aspiration level of 0.25
- s_1^- = the degree that the ratio between tables and chairs is under the aspiration level of 0.25
- s_2^+ = the amount of storage capacity overutilized when the aspiration level is 200,000 cubic feet
- s_2^- = the amount of storage capacity underutilized when the aspiration level is 200,000 cubic feet
- s_3^+ = the amount of inventory cost overspent when the aspiration level is $7,000
- s_3^- = the amount of inventory cost underspent when the aspiration level is $7,000

Before incorporating the deviational variables, the original NLP model must be modified to include the table/chair ratio constraint. Considering that the amount of tables to order is x_1 and the amount of chairs to order is x_2, the ratio constraint can be written as a nonlinear equation:

$$\frac{x_1}{x_2} = 0.25$$

The objective function seeks to minimize the negative and positive deviations for the ratio constraint, positive deviation for the capacity, and positive deviation for the inventory operating cost. The inventory operation cost is, in fact, the objective function in the original NLP model and is transformed into a constraint where the right-hand side value is $7,000. Thus, the overall nonlinear GP (NLGP) model can be formulated as:

$$\text{Minimize } Z = 1000(s_1^- + s_1^+) + 50s_2^+ + s_3^+$$

and is subject to:

$$\frac{x_1}{x_2} + s_1^- - s_1^+ = 0.25 \qquad \text{(priority 1: ratio constraint)}$$

$$\sum_{j=1}^{5} s_j x_j + s_2^- - s_2^+ = 200,000 \qquad \text{(priority 2: capacity)}$$

$$\sum_{j=1}^{5} \left(h_j \frac{x_j}{2} + o_j \frac{D_j}{x_j} \right) + s_3^- - s_3^+ = 7000 \qquad \begin{array}{l}\text{(priority 3: inventory}\\ \text{operating cost)}\end{array}$$

$$\sum_{j=1}^{5} \left(h_j \frac{x_j}{2} + o_j \frac{D}{x_j} + p_j D_j \right) \leq 1,500,000 \quad \text{(system constraint: budget)}$$

and

$x_j, s_1^+, s_1^-, s_2^+, s_2^-, s_3^+, s_3^- \geq 0$ for all $j=1, 2, 3, 4, 5$ (non-negativity requirements)

Solving NLGP Models with Solver

The previous GP model is expressed as a minimization NLP model with 11 decision variables (five functional and six deviational) and four constraints (three GP and one system constraint). The complete template and modeling details for this problem can be found in the *ch5_furniture* file, which can be downloaded from the companion website. Figure 5-3 indicates the setup template for this model. Note that the initial values for the decision variables are set equal to the EOQ values and the initial values for the deviational variables are set to zero. This initial set of values serves as a good starting point and allows the GRG algorithm to first search for possible solutions in the neighborhood of EOQ values. The table/chair ratio for this solution is 0.52 and the decision maker has made it the first priority to change this ratio to 0.25.

Figure 5-4 shows the complete Solver setup and the final solution. Cell N24 is selected as the Set Objective cell and the goal is to minimize selected deviational variables according to their assigned priorities. The variable cells include decision variables (B16:F16) and deviational variables (J16: K16, J19: K19, and J22: K22). There are four constraints in this model:

- G24 <= B4 (total inventory value must not exceed the inventory budget of $1.5 million)
- L16 = M16 (table/chair ratio constraint of exactly 0.25)
- L19 = M19 (warehouse capacity constraint of no more than 200,000 cubic feet)
- L22 = M22 (inventory operating cost for the week of no more than $7,000)

3	Warehouse capacity (cubic feet)	200,000			
4	Average inventory budget	$1,500,000			
5					

		Table	Chairs	Beds	Sofas	Bookcases
6						
7	Weekly demand (units)	1125	2750	3075	3075	750
8	Purchase price per unit	$45	$85	$125	$155	$125
9	Holding cost (per unit per week)	$2	$3	$3	$3	$4
10	Ordering cost (per order)	$100	$225	$135	$135	$100
11	Storage space required (cubic feet per unit)	84	106	140	70	100
12						
13	Calculations and results					

		Table	Chairs	Beds	Sofas	Bookcases	Totals
14							
15	Economic order quantity (EOQ)	335	642	526	526	194	
16	Optimal order quantity (decision variables)	335	642	526	526	194	
17	Average inventory	168	321	263	263	97	
18	Average number of orders per week	3.35	4.28	5.85	5.85	3.87	
19	Total supply available	1125	2750	3075	3075	750	
20	Maximum cubic feet storage required	28174	68080	73650	36825	19365	226,094
21	Ordering cost per week	$335	$963	$789	$789	$387	$3,264
22	Holding cost per week	$335	$963	$789	$789	$387	$3,264
23	Inventory operating cost per week	$671	$1,927	$1,578	$1,578	$775	$6,529
24	Total inventory cost (ordering+holding +purc	$51,296	$235,677	$385,953	$478,203	$94,525	$1,245,654

Goal programming section:

	Chair/Table Ratio	s_negative	s_positive	LHV	RHV	
15	Ratio					
16	0.522233 <---=B16/C16					
P1:	0.522233	0	0	0.52223	0.25	0 <---=SUMPRODUCT(J16:K16,J17:K17)
		s_negative	s_positive			
		1000	1000	0	200,000	0 <---=SUMPRODUCT(J19:K19,J20:K20)
P2:		s_negative	s_positive	226.094		
		0	50			
P3:		s_negative	s_positive			
		0	0	$6,529	7000	0 <---=SUMPRODUCT(J22:K22,J23:K23)
			10			
			Objective function:	0		<---=N16+N19+N22

Figure 5-3 Initial setup for the furniture goal programming model

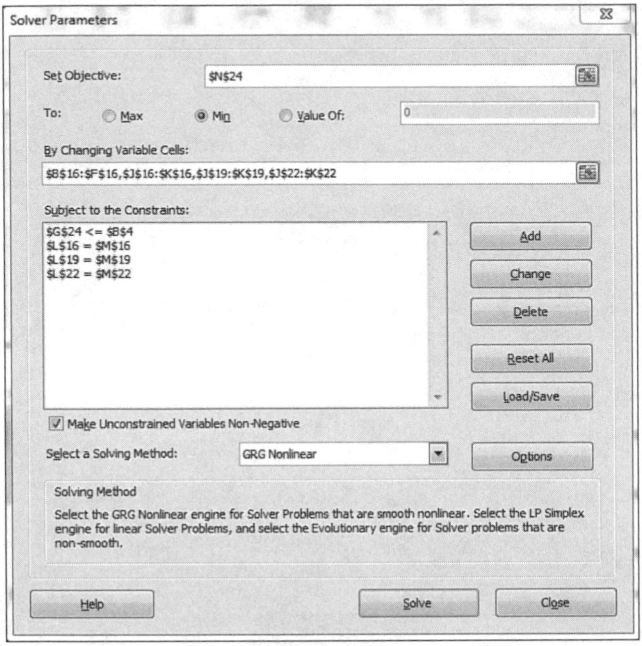

Figure 5-4 Solver Parameters setup for the furniture goal programming model

Figure 5-5 shows the final solution of the GP model. As shown, relaxing the constraint of the inventory operating cost allows the decision maker to completely meet the priorities. The warehouse can order 183 tables and 734 chairs and the ratio between tables and chairs is 0.25. In addition, the warehouse can order 279 beds, 526 sofas, and 194 bookcases every week. This ordering system will allow an additional space of 11,557 cubic feet and can be implemented within the inventory operating cost of $7,000 and with a total inventory value budget of $1,246,125, which is below the budget limit of $1.5 million.

		Table	Chairs	Beds	Sofas	Bookcases	Totals
3	Warehouse capacity (cubic feet)	200,000					
4	Average inventory budget	$1,500,000					
5							
6		Table	Chairs	Beds	Sofas	Bookcases	
7	Weekly demand (units)	1125	2750	3075	3075	750	
8	Purchase price per unit	$45	$85	$125	$155	$125	
9	Holding cost (per unit per week)	$2	$3	$3	$3	$4	
10	Ordering cost (per order)	$100	$225	$135	$135	$100	
11	Storage space required (cubic feet per unit)	84	106	140	70	100	
12							
13	Calculations and results						
14		Table	Chairs	Beds	Sofas	Bookcases	Totals
15	Economic order quantity (EOQ)	335	642	526	526	194	
16	Optimal order quantity (decision variables)	187	748	278	526	194	
17	Average inventory	93	374	139	263	97	
18	Average number of orders per week	6.02	3.68	11.05	5.85	3.87	
19	Total supply available	1125	2750	3075	3075	750	
20	Maximum cubic feet storage required	15702	79260	38953	36824	19365	190,105
21	Ordering cost per week	$602	$827	$1,492	$789	$387	$4,098
22	Holding cost per week	$187	$1,122	$417	$789	$387	$2,902
23	Inventory operating cost per week	$789	$1,949	$1,909	$1,578	$775	$7,000
24	Total inventory cost (ordering+holding +pur)	$51,414	$235,699	$386,284	$478,203	$94,525	$1,246,125

Chair/Table Ratio: 0.25 =B16/C16

	s_negative	s_positive	LHV	RHV	
P1:	0	0	0.25	0.25	0
	1000	1000			
P2:	9895.069	0	200,000	200,000	0
		50			
P3:	0	0	$7,000	7,000	0
		10			
Objective function					0

Figure 5-5 Final solution for the furniture goal programming model

Exploring Big Data with Goal Programming

In the era of Big Data, goal programming has the potential to become the favorite tool of choice for data analysts, as organizations try to meet multiple objectives under fierce competition. The *variety* of Big Data allows the decision maker to analyze business problems from many dimensions and multiple goals. Decision makers can combine data from internal and external sources and structured and unstructured formats to yield new insights into optimization models. A trucking company, for example, may use sensors to monitor fuel levels, driver behavior, container locations, and load capacity. These sensors produce data that can then be incorporated into logistical optimization algorithms with multiple goals: improving the efficiency of the route network, lowering the cost of fuel, and decreasing the risk of accidents.[50]

GP models can be formulated and solved as a series of connected programming models or as a single programming model. In the first approach, the model is designed as a linear or nonlinear LP model and is then solved with the first goal as the objective function. Once a solution is achieved, the objective function is transformed into a constraint where the right-hand side value is indeed the optimal value. In other words, the decision maker places the second goal as the objective function under the same original set of constraints with the added new requirement that the first goal continue to be satisfied. This process continues until all the goals are incorporated. The series of connected models approach is relatively simple and straightforward. However, when the decision maker is working with large problems and with many objectives, it becomes difficult to connect individual models. Often, objective functions (goals) contradict each other. This will most likely result in one or more models producing infeasible solutions.

Wrap Up

This chapter discussed formulation and solution methodologies for goal programming models. GP models are a special type of optimization technique, which seek to achieve multiple goals in a given order of priority. Today, companies generate and store large amounts of data. To remain competitive, companies make decisions based on multiple criteria and explore multiple opportunities, which in turn allow data scientists to simultaneously seek optimization of multiple measures. In such an environment, GP can become the data scientist's weapon of choice.

A more-efficient approach would be to design a single model that incorporates multiple goals. The chapter offers practical recommendations for transforming a regular linear or nonlinear programming model into a single model with ranked priorities. The step-by-step methodology for GP formulation is shown in Figure 5-6.

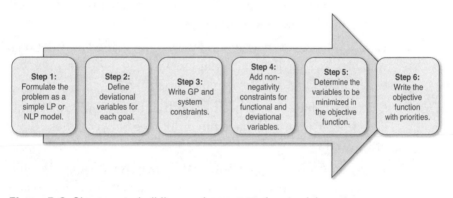

Figure 5-6 Six steps to building goal programming models

These steps can be used for both linear and nonlinear models. Once a model is designed, you can use Solver to achieve a solution via a simplex LP method for linear models or a GRG method for nonlinear functions. From a practical perspective, it is suggested that the GP solution start with a good initial solution in the trial template. You (the decision maker) may want to use the optimal solution of the original LP or NLP model as the initial search point.

The single model solution approach also allows you to test several what-if scenarios with different values for priorities. Using the allowable increase and decrease values provided by the Sensitivity Report, you can gain insights about the priority values and how much change is needed to affect the final solution.

Review Questions

1. What are goal programming models and how do they compare with other LP or NLP models?

2. Why are GP models important in business applications? Discuss advantages and disadvantages of GP as compared with other programming models.

3. Describe the process of formulating linear and nonlinear GP models. What are the steps involved and challenges faced by the decision maker?

4. What is the aspiration level and how can it be established? What is a goal deviation and how can it be established? What is a goal priority and how can it be established?

5. Decision variables in a GP model can be classified as functional variables and deviational variables. What is the difference between functional and deviational variables?

6. GP models have two types of constraints: systems constraints and GP constraints. What is the difference between these types of constraints?

7. Discuss the advantages of solving GP models with a single model that incorporates all the goals. How can the decision maker use sensitivity analysis to refine priority values and improve the model and its solution?

8. Nonlinear GP models are generally more difficult to solve than linear GP models. Explain why. What can the decision maker do to avoid such difficulties?

9. Explain the relationship between deviational variables and the attempt of the decision maker to under or overachieve an aspirational level. Why does minimizing the positive deviation generally lead to underachievement? Why does minimizing the negative deviation generally lead to overachievement?

Practice Problems

1. Consider the computer manufacturer case from Chapter 2 (problem 1). As a reminder, the following requirements are stated in the problem description:

   The Marketing Department projects that the expected demand for laptops will be at least 1,000, and for desktops it will be at least 800 per day. The production facility has a limited capacity of no more than 2,000 laptops and 1,700 desktops per day. The Sales Department indicates that contractual agreements of at most 2,000 computers per day must be satisfied. Each laptop computer sold results in $600 net profit and each desktop computer produces a $300 net profit.

   In addition, the decision maker wants to include the following goals with their respective priorities:

   - *Goal 1:* The company needs to fulfill the contractual agreement and produce 2,000 computers per day (P = 100).
   - *Goal 2:* The company should not produce more than the contractual agreement of 2,000 computers per day (P = 50).
   - *Goal 3:* The company should fully use the capacity of the production facility for each laptop and desktop (P = 500).
   - *Goal 4:* The company needs to make the maximum possible amount of net profit (P = 10).
     a. Formulate and solve the model as a linear programming model that seeks to determine how many desktops and laptops to produce to maximize the net profit.
     b. Formulate the problem as a goal programming model that seeks to meet multiple goals according to their priority. Identify goal constraints and system constraints. What is the optimal number of laptop and desktop computers to be made each day? Which of the goals are achieved?

2. Consider the case of the animal shelter LP model from Chapter 2 (problem 2). As a reminder, the following requirements are stated in the problem description:

You run an animal shelter facility and are asked to purchase the food supply for the next month. Purina costs $180 per bag. Each bag weighs 80 pounds and can feed as many as six animals per month. The Natural Choice bag costs $110, weighs 60 pounds, and can feed four animals for one month. You have been given a maximum budget of $4,000 for this purchase, although you don't have to spend that much. The facility has room for no more than 2,000 pounds and can board an average of 100 animals per day during a typical month.

a. Formulate the problem as a goal programming model to include the following goals with their respective priorities:

- *Goal 1:* The shelter needs to be able to feed 200 animals due to an unexpected increase in the shelter population (P = 1,000).

- *Goal 2:* The shelter needs to minimize the purchasing cost as much as possible (P = 50).

b. Solve the GP problems and indicate the optimal number of bags to be purchased for each type of food.

3. Consider the chair manufacturer from Chapter 3 (problem 1). As a reminder, the following requirements are stated in the problem description:

A chair furniture manufacturer makes different types of chairs, which are categorized into single-seat chairs or multiple-seat chairs. The manufacturer can make at most three times as many single-seat chairs as multiple-seat chairs. There are 5,000 machine hours and $15,000 production budget available each month. The operational data about production quantities, raw materials costs, total machine hours used, and total sales for each week for each chair are stored in the ch5_P3chairproduction file and can be downloaded from the companion website.

a. Formulate the problem as a goal programming model to include the following goals and with their respective priorities:

- *Goal 1:* Avoid the overutilization of 5,000 hours because it is hard to obtain in the labor market (P1 = 5).

- *Goal 2:* Meet the demand as much as possible (P2 = 10).

    b. Solve the GP problems and analyze the Answer and Sensitivity Reports. Offer managerial recommendations for the furniture manufacturer.

4. Consider the paint manufacturer case from Chapter 4 (problem 3). As a reminder, the following requirements are stated in the problem description:

Due to contractual agreements, there is a maximum production level of 1,000 units per each primer, and a minimum production level of 500 units. There are 10,000 machine hours available and $100,000 budget for raw materials. The net profit for each component is a function of the amount of each primer to be produced. Specifically, the net profit can be calculated as:

$$NP = \frac{\left(x_1^3 + x_2^3 + \ldots + x_n^3\right) - 3\left(x_1^2 + x_2^2 + \ldots + x_n^2\right) + 2\left(x_1 + x_2 + \ldots + x_n\right)}{10,000}$$

The file with operational data and the linear programming solution for this problem is named *ch5_P4paintprimer* and can be downloaded from the companion website at www.informit. com/title/9780133760354. Formulate and solve the problem as a goal programming model to include the following goals with their respective priorities:

- *Goal 1:* The company needs to make at least $800,000 in profit (P = 100).
- *Goal 2:* The company should not underutilize the available machine hours (P = 50).
- *Goal 3:* The company should not spend more than $80,000 in purchasing materials (P = 10).

5. Consider the toy manufacturing case from Chapter 4 (problem 4). As a reminder, the following requirements are stated in the problem description for this nonlinear programming model:

The manufacturing requirements for each toy production lot remain the same and are shown in the following table:

Raw Materials	Truck Toy	Car Toy	Available
Plastic	6 lb.	8 lb.	72 lb.
Labor hours	10 hrs.	8 hrs.	80 hrs.
Machine time	10 hrs.	4 hrs.	60 hrs.

The cost of producing a lot of toy trucks is $700T + 40T^2 + 1,000$ and the cost of producing a lot of toy cars is $200C + 20C^2 + 1,500$. There is a total budget of $5,000 per week. The profit for either toy is $500 per lot. The template and additional data can be found in the Excel file named ch5_P5toys *and can be downloaded from the companion website.*

a. Formulate the problem as a goal nonlinear programming model to include the following goals and with their respective priorities:

- *Goal 1:* The company should use the available plastic completely (P = 1,000).

- *Goal 2:* The company should not underutilize labor and machine hours available (P = 200).

- *Goal 3:* The company should minimize production cost (P = 500).

b. Solve the GP problems and indicate which of the goals is achieved. Try different priority values and investigate how the solution changes. Offer managerial recommendations based on your findings.

6. A shoe manufacturer would like to determine the best way to maximize profits on three new shoe models. The new models belong in the following categories: running, hiking, and casual. The model in the *running* category will result in a profit of $30 per item, the model in the *hiking* category will result in a $40 profit per item, and the model in the *casual* category will result in a $20 profit per item. The manufacturer has limitations in terms of production hours to consider: 1,000 hours of production time available per month in the Cutting/Sewing Department, 600 hours per month in the Finishing Department, and

only 500 hours per month available in the Packaging/Shipping Department. The following table illustrates the production time requirements for each shoe model:

Production Time (Hours)				
Model	**Cutting/ Sewing**	**Finishing**	**Packaging/ Shipping**	**Profit/ Shoe**
Running shoe model (x_1)	2.00	0.50	1.5	$30.00
Hiking shoe model (x_2)	3.00	1.50	0.125	$40.00
Casual shoe model (x_3)	1.00	0.50	1.5	$20.00

The production manager is seeking to optimize production according to several company goals:

- *(P1 = 500) Priority 1:* The manufacturer should produce at least 100 pairs of shoes for each model.
- *(P2 = 400) Priority 2:* The manufacturer should meet monthly demand of 300 pairs per month for running shoes and casual shoes.
- *(P3 = 100) Priority 3:* The manufacturer should utilize the available machine hours.

  a. Formulate the problem as a linear programming model.
  b. Define deviational variables for the goal programming model.
  c. Identify GP and system constraints.
  d. Formulate the GP model.
  e. Solve the problem with Solver and analyze the results.

7. The management team of a resort hotel and convention center has several goals related to the new project of expanding the convention center. You are hired as an analyst and your task is to help determine how many rooms are needed to expand. The following table indicates a summary of operational data, which can serve as input parameters to the optimization model:

Room Type	Required Area per Room (Square Feet)	Expected Cost per Room
Small rooms	1,400	$20,000
Medium rooms	1,800	$30,000
Large rooms	2,200	$40,000

There is a $2,000,000 budget for expansion and the following priorities must be considered:

- *Goal 1:* The expansion should include at least 5 small conference rooms (P1 = 5).

- *Goal 2:* The expansion should include at least 10 medium conference rooms (P2 = 3).

- *Goal 3:* The expansion should include at least 15 large conference rooms (P3 = 3).

- *Goal 4:* The expansion should include at least 90,000 square feet (P4 = 1).

- *Goal 5:* The expansion should cost no more than $2,000,000 (P5 = 1).

  a. Formulate the problem as a simple LP model. The goal is to minimize the overall construction cost and the constraints can be similar to the five stated goals with equal priority values.

  b. Define deviational variables, define goal and systems constraints, and reformulate the problem as a goal programming model.

  c. Solve the problem with Solver and analyze the results.

6

Business Analytics with Integer Programming

Chapter Objectives

- Discuss the need to require integer or binary values for the solution of some programming models
- Offer a graphical explanation of the integer programming models
- Discuss different types of integer programming models and when to choose them
- Demonstrate the process of seeking integer or binary solutions for linear, nonlinear, or goal programming models via Solver
- Discuss the main assumptions of the knapsack and assignment problems
- Describe the challenges of requiring binary, integer, or mixed integer solutions for programming models
- Offer practical recommendations when using integer, binary, or mixed programming models in the era of Big Data

Prescriptive Analytics in Action: Zara Uses Mixed IP Modeling[1]

Zara is a Spanish clothing and accessories retailer founded in 1975 and is considered to be one of the largest international fashion companies with over 2,300 stores around the world.[2] Unlike other clothing stores, Zara has chosen to vertically integrate its supply chain. This allows the company to replenish inventory directly to every Zara store in the world twice a week. Determining the exact number of units of each size for each article that should be in each shipment to each store is a challenging task. There are approximately several million shipment decisions to be made and these decisions require an enormous amount of operational data.

Further, the decision must be made within a few hours and each decision must also consider the limitation of the available inventory in the warehouses. Data is captured in Zara stores using both personal digital assistant (PDA) devices and point of sale (POS) transaction processing systems. The customer preference data on the PDAs is linked to the customer purchase information data and uploaded for analysis. Each store also has a POS transaction processing system that records customer purchase information. This data allows Zara to use advanced analytical tools to generate business intelligence and allow the store managers, designers, and suppliers to make quick decisions. For example, recently Zara's analytics team developed a large-scale, mixed integer programming model in addition to other methods, such as forecasting algorithms and stochastic analysis. The project was implemented in all the stores with tremendous results. The seasonal sales increased over 3% to 4% and reduced the transshipment cost significantly. The sales resulted in a financial benefit between $233 million and $353 million per year, and the net income increased between $28 million and $42.4 million.

---

[1] https://www.informs.org/Sites/Getting-Started-With-Analytics/Analytics-Success-Stories/Case-Studies/Zara.

[2] http://www.inditex.com/en/who_we_are/stores.

Introduction

The programming models discussed so far are based on the assumptions of *divisibility*. Divisibility allows decision variables to take integer as well as fractional values. This is an acceptable and often reasonable assumption in many applications. For example, a company can produce 9.4 gallons of a divisible good such as milk. Also, it is safe to accept an order quantity of 189.2 tables as an optimal ordering quantity. In practical terms, the order can be rounded to 189 tables.

Often, the impact on the objective function of the rounding of a fractional solution into an integer solution is often minimal. However, there are business applications where the solutions must be restricted to be an integer. Assume that a construction company wants to determine the number of houses and apartment buildings to construct in a new subdivision. A solution suggesting that the company should build 12.5 houses and 5.3 apartment buildings may be optimal, but not practical. The company cannot build 0.5 houses or 0.3 buildings. The question becomes: Should the company round the number of houses to 12 or 13 houses or the number of buildings to 5 or 6? Do any of these roundings still offer a feasible solution, or is there another combination of integer values for houses and buildings that provides a better value for the objective function?

Integer programming (IP) models seek optimal solutions, while all or some of the decision variables are required to be integers. IP models have the same structure as the LP, NLP, or GP models discussed in the previous chapters. They have an objective function to be optimized and a set of constraints to be satisfied. In addition, IP models have a set of constraints that forces some or all decision variables to be integers.

Formulation and Graphical Solution of IP Models

The following is the formulation of an integer linear programming model with two decision variables:

$$\text{Max } Z = 400x_1 + 300x_2 \quad \text{(objective function)}$$

subject to:

$$10x_1 + 15x_2 \le 150 \qquad \text{(constraint 1)}$$

$$x_1 \ge 3 \qquad \text{(constraint 2)}$$

$$x_2 \ge 3 \qquad \text{(constraint 3)}$$

$$x_1, x_2 \ge 0 \text{ and integer} \qquad \text{(non-negativity and integer constraints)}$$

If the phrase "and integer" is removed from the preceding formulation, then a typical LP model is presented. In fact, this is the formulation discussed in Chapter 2, "Introduction to Linear Programming," for the Rolls Bakery problem, where x_1 represents the number of lots for dinner roll cases (DRCs) and x_2 represents the number of lots for sandwich roll cases (SRCs). The only difference is the change of constraint 3 from $x_2 \ge 4$ to $x_2 \ge 3$. It is assumed that the market demand for SRCs has dropped to 3,000 cases, and because each lot has 1,000 cases, the bakery may run three or more lots of SRCs. Recall that the optimal solution for the problem indicated that the bakery should run nine lots of DRCs and four lots of SRCs for a total net profit of $4,800. However, because the minimum requirement for SRCs has changed, the new solution allows for making 10.5 lots of DRCs and three lots of SRCs for a total net profit of $5,100, as shown graphically in Figure 6-1a. If the bakery can only run full lots during the week, then the decision maker must seek integer solutions. Figure 6-1b indicates all the points in the area of feasible solutions where the values of x_1 and x_2 are integers. When moving the line of the objective function to the extreme point, the optimal solution is achieved at ten lots of DRCs and three lots of SRCs for a total net profit of $4,900.

As shown in the graphical representation, adding the integer constraints in the regular LP models causes a significant change in the nature of the problem. While the regular LP models have an infinite number of possible solutions all within the area of feasible solutions, the integer counterpart has a limited number of possible solutions, represented only by the integer points within the area of feasible solutions (27 in the case of Rolls Bakery problem).

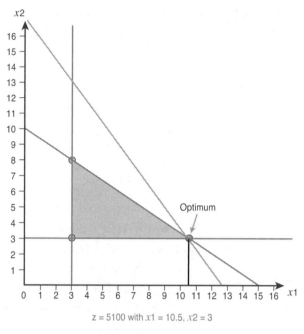

z = 5100 with $x1 = 10.5$, $x2 = 3$

(a) Non-integer solution

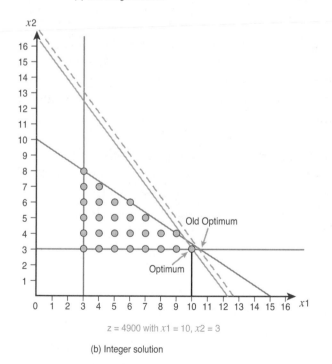

z = 4900 with $x1 = 10$, $x2 = 3$

(b) Integer solution

Figure 6-1 Graphical solution of Rolls Bakery IP model

The finite nature of IP models leads to simpler solution algorithms for IP models. Solver, for example, uses an algorithm that systematically checks for all integer points within the area of feasible solutions and returns the one that offers the best value for the objective function. This is seemingly a daunting task considering the potential large number of integer points in a real-life application, but Solver can treat each point as a separate LP model and execute it swiftly.

Types of Integer Programming Models

When *all* decision variables are required to be integers, the model is known as an *all-integer programming* model. When some, but not all, variables are required to be integers, the model is known as a *mixed-integer programming* model. As mentioned earlier, the IP models with linear constraints and objective function are called linear IP models. When the objective function or at least a constraint is not linear, the model is known as a nonlinear IP model.

Nonlinear IP models offer special challenges for solution algorithms. As mentioned in Chapter 4, "Business Analytics with Nonlinear Programming," the values of *reduced gradients* and *Lagrange multipliers* are valid only at the point of the optimal solution. So, although the dual values may locate integer values around the optimal point, there is no range within which to locate all possible integer values. The generalized reduced gradient (GRG) algorithm is unable to distinguish between local and global optima. Searching for the optimal integer solution becomes even more difficult for nonsmooth, nonlinear models. Usually, nonsmooth relationships can be modeled with *if-then* or *lookup* functions in Excel, such as VLOOKUP() or HLOOKUP(). In these situations, the evolutionary solving method of Solver is suggested.

Sometimes, decision variables are required to take binary values. For example, instead of deciding how many houses to build, a construction company may need to decide whether to build or not build houses. Instead of calculating how many units of chairs to order, the warehouse may need to decide whether to order or not order chairs in the next shipment. These requirements can be modeled with a special case of integer programming called *binary integer programming* or simply *0-1 programming*.

Solving Integer LP Models with Solver

Figure 6-2 shows the final values of the noninteger solution of the adjusted Rolls Bakery problem. As noted graphically, the x_1 = 10.5 and x_2 = 3 for a total net profit of $5,100. The Add Constraint window is used to enforce integer values for the decision variables located in cells B8 and B9. The Simplex LP method is used to search for the optimal solution, as indicated in the Solver Parameters dialog box shown in Figure 6-3.

	A	B	C	D	E	F	G	H
1				Rolls Bakery Production Information				
2	Products	Wholesale Price per Case	Wholesale Price per Lot	Processing Time (in hours) per Lot	Cost of Raw Materials per Lot	Net Profit per Lot	Demand for Cases	Demand for Production Lots
3	Dinner Roll Case (DRC)	$0.75	$750	10	$250	$400	3000	3
4	Sandwich Roll Case (SRC)	$0.65	$650	15	$200	$300	3000	3
5								
6			What-if Analysis					
7	Functional Variables	Number of Lots per Week	Net Profit Per Lot	Total Machine Time Used (in hours)	Right hand-side values			
8	DRC Lots (x1)	10.5	$4,200	105	3	Minimum DRC (Constraint 2)		
9	SRC Lots (x2)	3	$900	45	3	Minimum SRC (Constraint 3)		
10			$5,100	150	150	Machine Hours (Constraint 1)		
11		Add Constraint						
12								
13		Cell Reference:			Constraint:			
14		$B$8:$B$9		int	integer			
15				<=				
16				=				
17		OK		>=	int	Cancel		
18				bin dif				

Figure 6-2 Adding integer constraints to the Rolls Bakery problem

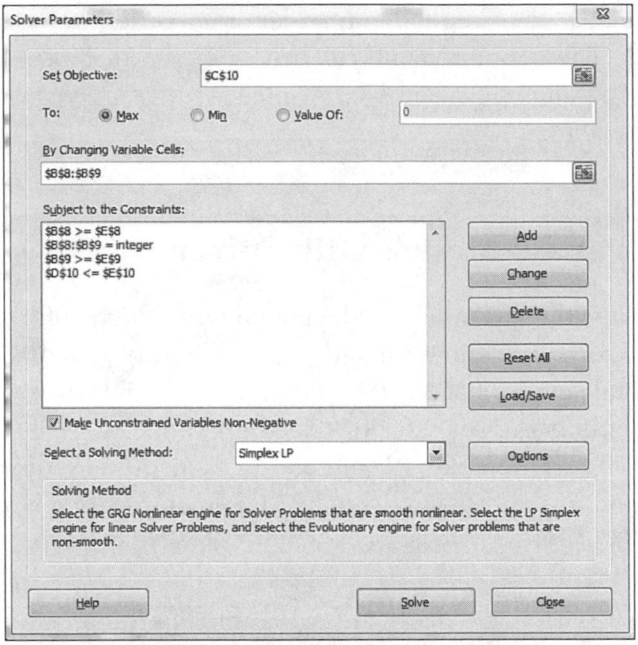

Figure 6-3 Solver Parameters dialog box for the Rolls Bakery integer LP model

There are two results you should anticipate when solving integer programming models. First, the value of the objective function will not be better than the objective function of the noninteger model. The integer requirement only adds more constraints to the original set of constraints. As such, the area of feasible solution is only limited to the integer points (as shown graphically), and the solution can only get worse. Second, the values of decision variables will no longer have fractions. As shown in Figure 6-4, the Z value in cell C10 is $4,900, a reduction in the net profit from the original solution of $5,100. Also, the final values for the decision variables are integers 10 and 3, as compared with the previous values of 10.5 and 3.

	A	B	C	D	E	F	G	H
	Products	Wholesale Price per Case	Wholesale Price per Lot	Processing Time (in hours) per Lot	Cost of Raw Materials per Lot	Net Profit per Lot	Demand for Cases	Demand for Production Lots
3	Dinner Roll Case (DRC)	$0.75	$750	10	$250	$400	3000	3
4	Sandwich Roll Case (SRC)	$0.65	$650	15	$200	$300	3000	3
5								
6		**What-If Analysis**						
	Functional Variables	Number of Lots per Week	Net Profit Per Lot	Total Machine Time Used (in hours)	Right hand-side values			
8	DRC Lots (x1)	10	$4,000	100		3	Minimum DRC (Constraint 2)	
9	SRC Lots (x2)	3	$900	45		3	Minimum SRC (Constraint 3)	
10			$4,900	145		150	Machine Hours (Constraint 1)	

Figure 6-4 Final solution values for the integer LP model

Solving Nonlinear IP Models with Solver

Chapter 4 discussed the WCF's warehouse inventory problem. Recall that this is a nonlinear model. The following is the integer NLP formulation:

$$\text{Minimize } Z = \sum_{j=1}^{5} \left(h_j \frac{x_j}{2} + o_j \frac{D_j}{x_j} \right) \quad \text{(nolinear objective function)}$$

subject to:

$$\sum_{j=1}^{5} s_j x_j \leq C \quad \text{(linear constraint)}$$

$$\sum_{j=1}^{5} \left(h_j \frac{x_j}{2} + o_j \frac{D}{x_j} + p_j D_j \right) \leq P \quad \text{(nonlinear constraint)}$$

$x_j \geq 0$ and integers for all $j=1, 2, 3, 4, 5$ (non-negativity and integer constraints)

Figure 6-5 and Figure 6-6 show the noninteger and integer solutions to the preceding problem. As shown, the decision variables have changed from 288.69, 574.93, 457.21, 469.20, and 179.53 to 291.00, 570.00, 452.00, 486.00, and 178.00. Note that the decision variables are not simply rounded to generate the optimal solution. The value of the objective function is an interesting observation: The objective function is slightly better than the original value, from $6,576 to $6,575, contrary to the expectation discussed earlier in the case of integer LP models. This is an indication of particularities in the behavior of nonlinear models. It is quite possible, as discussed in the "Challenges to NLP Models" section of Chapter 4, that the GRG algorithm may not have produced the optimal solution when trying to solve the original NLP model. A better solution is found now, when the initial values of decision variables have a good starting point, as indicated in Figure 6-7.

	A	B	C	D	E	F	G
3	Warehouse Capacity (cubic feet)	200,000					
4	Inventory Budget	$1,500,000					
5							
6		Tables	Chairs	Beds	Sofas	Bookcases	
7	Weekly Demand (units)	1125	2750	3075	3075	750	
8	Purchase Price per Unit	$45	$85	$125	$155	$125	
9	Holding Cost (per unit, period)	$2	$3	$3	$3	$4	
10	Ordering Cost (per order)	$100	$225	$135	$135	$100	
11	Storage Space Required (cubic feet per unit)	84	106	140	70	100	
12							
13	Calculations and Results						
14		Tables	Chairs	Beds	Sofas	Bookcases	Totals
15	Economic Order Quantity (EOQ)	335	642	526	526	194	
16	Optimized Order Quantity	288.69	574.93	457.21	469.20	179.53	
17	Average Inventory	144	287	229	235	90	
18	Average Number of Orders per Week	3.90	4.78	6.73	6.55	4.18	
19	Total Supply Available	1125	2750	3075	3075	750	
20	Maximum Cubic Foot Storage Required	24250	60942	64010	32844	17953	200000
21	Ordering Cost per Week	$390	$1,076	$908	$885	$418	$3,676
22	Holding Cost per Week	$289	$862	$686	$704	$359	$2,900
23	Inventory Operating Cost per Week	$678	$1,939	$1,594	$1,589	$777	$6,576
24	Total Inventory Value	$51,303	$235,689	$385,969	$478,214	$94,527	$1,245,701

Figure 6-5 Noninteger solutions for the WCF inventory problem

	A	B	C	D	E	F	G
3	Warehouse Capacity (cubic feet)	200,000					
4	Inventory Budget	$1,500,000					
5							
6		Tables	Chairs	Beds	Sofas	Bookcases	
7	Weekly Demand (units)	1125	2750	3075	3075	750	
8	Purchase Price per Unit	$45	$85	$125	$155	$125	
9	Holding Cost (per unit, period)	$2	$3	$3	$3	$4	
10	Ordering Cost (per order)	$100	$225	$135	$135	$100	
11	Storage Space Required (cubic feet per unit)	84	106	140	70	100	
12							
13	Calculations and Results						
14		Tables	Chairs	Beds	Sofas	Bookcases	Totals
15	Economic Order Quantity (EOQ)	335	642	526	526	194	
16	Optimized Order Quantity	291.00	570.00	452.00	486.00	178.00	
17	Average Inventory	146	285	226	243	89	
18	Average Number of Orders per Week	3.87	4.82	6.80	6.33	4.21	
19	Total Supply Available	1125	2750	3075	3075	750	
20	Maximum Cubic Foot Storage Required	24444	60420	63280	34020	17800	199964
21	Ordering Cost per Week	$387	$1,086	$918	$854	$421	$3,666
22	Holding Cost per Week	$291	$855	$678	$729	$356	$2,909
23	Inventory Operating Cost per Week	$678	$1,941	$1,596	$1,583	$777	$6,575
24	Total Inventory Value	$51,303	$235,691	$385,971	$478,208	$94,627	$1,245,700

Figure 6-6 Integer solutions for the WCF inventory problem

19	Variable Cells			
20	Cell	Name	Original Value	Final Value
21	$B$16	Optimized Order Quantity Tables	288.69	291.00
22	$C$16	Optimized Order Quantity Chairs	574.93	570.00
23	$D$16	Optimized Order Quantity Beds	457.21	452.00
24	$E$16	Optimized Order Quantity Sofas	469.20	486.00
25	$F$16	Optimized Order Quantity Bookcases	179.53	178.00

Figure 6-7 Change in the decision variables as indicated in the Answer Report

Solving Integer GP Models with Solver

The Rolls Bakery production problem was also discussed in Chapter 5, "Business Analytics with Goal Programming," as a goal programming model. Adding the integer requirement to the GP formulation shown in Chapter 5, the integer GP becomes:

$$\text{Minimize } Z = 300s_4^+ + 300s_5^+ + 300s_2^- + 300s_3^- + 20s_1^+ + 10s_6^- \text{ (GP objective function)}$$

subject to:

$$10x_1 + 15x_2 + s_1^- - s_1^+ = 150 \qquad \text{(machine hours)}$$

$$x_1 + s_2^- - s_2^+ = 3 \qquad \text{(minimum demand for DRC)}$$

$$x_2 + s_3^- - s_3^+ = 4 \qquad \text{(minimum demand for SRC)}$$

$$x_1 + s_4^- - s_4^+ = 5 \qquad \text{(maximum demand for DRC)}$$

$$x_2 + s_5^- - s_5^+ = 6 \qquad \text{(maximum demand for SRC)}$$

$$400x_1 + 300x_2 + s_6^- - s_6^+ = 4800 \quad \text{(net profit constraint)}$$

and

$$x_1, x_2, s_1^-, s_1^+, s_2^-, s_2^+, s_3^-, s_3^+, s_4^-, s_4^+, s_5^-, s_5^+, s_6^-, s_6^+ \geq 0 \text{ and } x_1, x_2$$
are integers

Figure 6-8 shows the modification of the original GP model to add the integer constraints. Figure 6-9 shows the Answer Report for the integer solutions to the preceding problem. As shown, the decision variables have changed from 7.5 and 6 to 9 and 4, respectively. Both solutions yield to the same overall net profit ($4,800), but the integer solution meets the priorities slightly differently than the non-integer solution. As expected, the final value of the goal programming objective function has increased from 1050 to 1200.

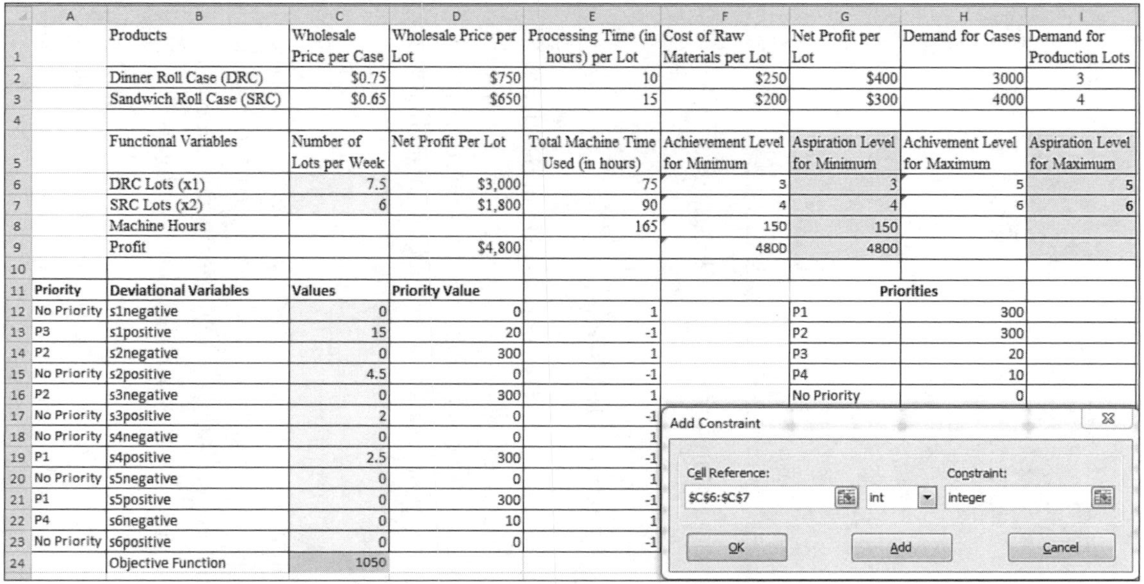

	A	B	C	D	E	F	G	H	I
1		Products	Wholesale Price per Case	Wholesale Price per Lot	Processing Time (in hours) per Lot	Cost of Raw Materials per Lot	Net Profit per Lot	Demand for Cases	Demand for Production Lots
2		Dinner Roll Case (DRC)	$0.75	$750	10	$250	$400	3000	3
3		Sandwich Roll Case (SRC)	$0.65	$650	15	$200	$300	4000	4
4									
5		Functional Variables	Number of Lots per Week	Net Profit Per Lot	Total Machine Time Used (in hours)	Achievement Level for Minimum	Aspiration Level for Minimum	Achivement Level for Maximum	Aspiration Level for Maximum
6		DRC Lots (x1)	7.5	$3,000	75	3	3	5	5
7		SRC Lots (x2)	6	$1,800	90	4	4	6	6
8		Machine Hours			165	150	150		
9		Profit		$4,800		4800	4800		
10									
11	Priority	Deviational Variables	Values	Priority Value			Priorities		
12	No Priority	s1negative	0	0	1		P1	300	
13	P3	s1positive	15	20	-1		P2	300	
14	P2	s2negative	0	300	1		P3	20	
15	No Priority	s2positive	4.5	0	-1		P4	10	
16	P2	s3negative	0	300	1		No Priority	0	
17	No Priority	s3positive	2	0	-1				
18	No Priority	s4negative	0	0	1				
19	P1	s4positive	2.5	300	-1				
20	No Priority	s5negative	0	0	1				
21	P1	s5positive	0	300	-1				
22	P4	s6negative	0	10	1				
23	No Priority	s6positive	0	0	-1				
24		Objective Function	1050						

Add Constraint
Cell Reference:
$C$6:$C$7
int
Constraint:
integer
OK
Add
Cancel

Figure 6-8 Enforcing integer solution to the Rolls Bakery problem

14	Objective Cell (Min)			
15	Cell	Name	Original Value	Final Value
16	$C$24	Objective Function Values	1050	1200
17				
18				
19	Variable Cells			
20	Cell	Name	Original Value	Final Value	Integer
21	$C$6	DRC Lots (x1) Number of Lots per Week	7.5	9	Integer
22	$C$7	SRC Lots (x2) Number of Lots per Week	6	4	Integer
23	$C$12	s1negative Values	0	0	Contin
24	$C$13	s1positive Values	15	0	Contin
25	$C$14	s2negative Values	0	0	Contin
26	$C$15	s2positive Values	4.5	6	Contin
27	$C$16	s3negative Values	0	0	Contin
28	$C$17	s3positive Values	2	0	Contin
29	$C$18	s4negative Values	0	0	Contin
30	$C$19	s4positive Values	2.5	4	Contin
31	$C$20	s5negative Values	0	2	Contin
32	$C$21	s5positive Values	0	0	Contin
33	$C$22	s6negative Values	0	0	Contin
34	$C$23	s6positive Values	0	0	Contin

Figure 6-9 Answer Report for the integer GP solution

The Assignment Method

The assignment method is a popular IP model that refers to assigning resources to a specific task. Only one resource can be assigned in a task, and only one task can be assigned to each resource. The goal is to maximize the revenue or minimize the cost of such assignments. This context can be generalized in many business problems:

- There are m machines and n workers to be assigned at the beginning of each shift. Each worker can be assigned to any machine; however, the productivity of each worker varies for each machine. The shift manager wants to assign each worker to each machine to maximize shift productivity.

- Suppose that a transportation company has three trucks available and three loads to be picked up as soon as possible. The company prides itself on timely response and delivery, so for each truck, the "cost" of loading the cargo will depend on the time taken for the truck to reach the pickup location. The shipment manager wants to find the best combination of truck and loads that results in the least total cost.

- A construction company has created several crews to be assigned to work in several construction projects. The crew can work in any project. Due to the set of skills allocated to each crew, the time to complete a project by each crew varies significantly. The company wants to minimize the total time needed to complete all projects.

General Formulation of the Assignment Problem

If the decision variables for the assignment problem are defined as:

$$x_{ij} = \begin{cases} 1 \text{ if item } i \text{ is assigned to project } j \\ 0 \text{ otherwise} \end{cases}$$

Then the problem can be formulated as:

$$\text{Max (or Minimize) } Z = \sum_{i=1}^{m}\sum_{j=1}^{n} v_{ij}x_{ij}$$

subject to:

$$\sum_{i=1}^{m} x_{ij} = 1 \quad \text{for all } j = 1,2,3, \dots ,n$$

$$\sum_{j=1}^{n} x_{ij} = 1 \quad \text{for all } i = 1,2,3, \dots ,m$$

x_{ij} are binary

where v_{ij} is the contribution coefficient of assigning item i to project j.

Solving the Assignment Method with Solver

A dispatcher at a trucking company has 14 trucks, which have just unloaded their shipment and are waiting to be repositioned for the next load. The trucks are currently located in the cities as shown in Table 6-1 and are indicated as unload cities. The dispatcher wants each truck to travel to the other cities where eight loads are waiting to be picked up. Some of the cities are repeated because there is more than one truck waiting in that specific city or more than one load waiting to be picked up. Each truck can transport only one load at a time. Also, not all trucks will be assigned because there are more trucks than loads available. Where should the dispatcher send each truck to minimize the total transportation distance of repositioning trucks from their unload city to the pickup (load) city?

Considering the general formulation discussed earlier in the chapter, the decision variables for the asxsignment problem are defined as:

$$x_{ij} = \begin{cases} 1 \text{ if truck } i \text{ is assigned to load city } j \\ 0 \text{ otherwise} \end{cases}$$

where $i = 1,2,3, \dots ,14$ and $j = 1,2,3, \dots ,8$

Then the repositioning distance Z can be calculated as:

Minimize $Z = 927x_{11} + 1505x_{12} + 1505x_{13} + \dots + 1190x_{18} +$

$+ 927x_{21} + 1505x_{22} + 1505x_{23} + \dots + 1190x_{28} +$

$+ 0x_{31} + 578x_{32} + 578x_{33} + \dots + 262x_{38} +$

$\dots$

$+ 1539x_{14,1} + 2022x_{14,2} + 2022x_{14,3} + \dots + 1753x_{14,8} +$

There are 14 trucks available to pick up eight loads. As a result, not all trucks will be assigned to a load, and no more than one truck will be assigned to a load. This requirement is enforced with the following set of equations:

$$x_{11} + x_{12} + x_{13} + \dots + x_{18} \leq 1$$

$$x_{21} + x_{22} + x_{23} + \dots + x_{28} \leq 1$$

$$x_{31} + x_{32} + x_{33} + \dots + x_{38} \leq 1$$

$$\dots$$

$$x_{14,1} + x_{14,2} + x_{14,3} + \dots + x_{14,8} \leq 1$$

Table 6-1 Distances from the Unload to Load Cities

From\To	Baltimore	Boston	Boston	Chicago	Miami	New Orleans	New York	Newark
Atlanta	927	1505	1505	944	974	682	1200	1190
Atlanta	927	1505	1505	944	974	682	1200	1190
Baltimore	0	578	578	973	1539	1607	272	262
Boston	578	0	0	1367	2022	2184	306	315
Boston	578	0	0	1367	2022	2184	306	315
Chicago	973	1367	1367	0	1912	1340	1145	1131
Chicago	973	1367	1367	0	1912	1340	1145	1131
Denver	2422	2839	2839	1474	2773	1737	2617	2602
Denver	2422	2839	2839	1474	2773	1737	2617	2602
Indianapolis	819	1295	1295	263	1651	1147	1035	1021
Jacksonville	1096	1636	1636	1387	526	810	1344	1338
Memphis	1273	1824	1824	773	1404	577	1533	1520
Memphis	1273	1824	1824	773	1404	577	1533	1520
Miami	1539	2022	2022	1912	0	1075	1756	1753

At the same time, at least one truck must be assigned to a load, and a load must be assigned to a truck. This requirement is represented with the following set of equations:

$$x_{11} + x_{21} + x_{31} + \ldots x_{14,1} = 1$$

$$x_{12} + x_{22} + x_{32} + \ldots + x_{14,2} = 1$$

$$x_{13} + x_{23} + x_{33} + \ldots + x_{14,3} = 1$$

$$\ldots$$

$$x_{18} + x_{28} + x_{38} + \ldots + x_{14,8} = 1$$

Finally, all decision variables must be either 0s or 1s:

$$x_{11}, x_{12}, x_{13}, \ldots, x_{14,6} + x_{14,7} + x_{14,8} \text{ are binary}$$

Figure 6-10 shows the Solver formulation and the solution of this assignment problem. The complete data and solution can be found in the *ch6_assignment* file, which can be downloaded from the companion website. The decision variables are located in cells B2:I15. Cells B19:I32 hold the repositioning distance from each origin/unload city to the pickup/load city. Each cell is the product of distance between the two respective cities and the decision variables. For example, cell B19 is calculated as:

B19 = B2 × (distance between Atlanta and Baltimore)

Note that cell B2 is binary, so the repositioning distance is 0 when B2 = 2 or 927 (distance between Atlanta and Baltimore). The repositioning distance from each city is summed in column J (cells J19:J32) and the overall sum of repositioning distance in shown in cell J33. This is, in fact, the objective function (value Z) and is placed in the Set Objective field in the Solver Parameters dialog box. The decision variables B2:I15 are placed in the By Changing Variable Cells field.

Figure 6-10 Solver formulation and solution for the repositioning problem

The first constraint in the Subject to the Constraints section enforces the requirement that the decision maker must assign at least one truck to each load. The second constraint forces the decision variables to be binary values, and the third constraint ensures that although not all trucks are assigned to a load, the decision maker does not have more than one truck available from the origin/unload city. This is a linear programming model, and as such the Simplex LP method is selected.

The solution indicates that the following assignment must be made:

1. The Baltimore load should be picked up by a truck in Baltimore.
2. The Boston load should be picked up by a truck in Boston.
3. The other Boston load should be picked up by a truck in Boston.
4. The Chicago load should be picked up by a truck in Chicago.
5. The Miami load should be picked up by a truck in Miami.
6. The New Orleans load should be picked up by a truck in Memphis.
7. The New York load should be picked up by a truck in Indianapolis.
8. The Newark load should be picked up by a truck in Chicago.

Two trucks in Atlanta, two trucks in Denver, one truck in Jacksonville, and one truck in Memphis are not assigned to pick up a load. The minimum total repositioning distance is 2,743 miles.

The number of decision variables in the assignment problem tends to grow significantly. The repositioning problem had 14 trucks and 8 load cities for a total of 14×8 or 112 binary variables. Imagine that there are 50 trucks and 50 loads; the number of decision variables becomes 2,500. Note that Solver has a limit of 200 decision variables. The Premium Solver offered by Frontline can extend this limit to 2,000 decision variables for LP models and 500 decision variables for NLP models.

The Knapsack Problem

The knapsack problem is a famous IP model that refers to a hiker deciding to select the most valuable items to carry in a hiking venture considering a weight limit. This context can be generalized in many business problems:

- There are many candidate projects that a company may select. Each project is characterized by a total budget and a given return on investment (ROI) rate. Considering a maximum allowable spending limit, the company needs to decide which projects to select to maximize the overall ROI.

- A fund manager is considering several potential investments and has estimated the return expected from each one. Each investment has a certain cost and the manager may not exceed a given budget. Which investments should the manager choose?

- A shipping company is considering shipping manufactured products from China to the United States. The volume of the shipping container is relatively small, and the company can only load a limited number of items.

- A carpet company produces huge rolls of carpet; the dimensions are determined by the manufacturing process. The rolls are later divided into smaller rolls according to customer orders. The manufacturers must decide how to cut lengths of carpet to satisfy customer orders while extracting the greatest value from each carpet.

General Formulation of the Knapsack Problem

If the decision variables for the knapsack problem are defined as:

- x_j—The number of items of type j to be included in the knapsack for each of the n potential items, then the problem can be formulated as:

$$\text{Maximize } Z = \sum_{j=1}^{n} r_j x_j$$

subject to:

$$\sum_{j=1}^{n} c_j x_j \leq C$$

$x_j \geq 0$ and integer

where:

c_j = the cost or weight of each type-j item, for j = 1, 2, ..., n

r_j = the reward associated with each type-j item, for j = 1, 2, 3, ..., n

C = the total budget or capacity of the knapsack

Adding integer constraints in the primer production problem discussed in Chapter 3, "Business Analytics with Linear Programming," transforms the LP formulation model into a knapsack problem. Recall that the company needed to determine what type of paint primer to produce and at what quantity in order to maximize the profit $(x_1, x_2, x_3, ..., x_{48})$ among its 48 types of paint primer. There was a limited number of machine hours (C = 4,000) and each type of primer generated a different amount of revenue $(r_1, r_2, r_3, ..., r_{48})$. Also, each primer uses a certain amount of machine hours $(c_1, c_2, c_3, ..., c_{48})$. The solution methodology with Solver is very similar to the one discussed in Chapter 3. The only difference would be adding a new set of constraints requiring the decision variables to be integers.

Exploring Big Data with Integer Programming

With respect to solution algorithms used by Solver, the IP models are quite different from the regular problems. Linear IP models have a finite number of possible solutions and can be found relatively fast with Solver. The nonlinear IP models require a more complicated algorithm to reach an optimal solution, and the likelihood that the solution is a local optimum is very high.

Adding integer or binary requirements to any model will result in a value of the objective function, which is equal or worse than the noninteger models. Although the assignment problem is very common, it often provides computational difficulty when using Solver. The number of decision variables in the assignment problem tends to grow significantly when the number of resources or tasks increases. To deal with such complexity, many decision makers are using new and improved Solver algorithms. New software programs, such as MATLAB,[51] XPRESS,[52] CPLEX,[53] and Gurobi,[54] have added integer solvers into their optimization suites. While integer programming algorithms continue to improve, formulating integer programming models is still an art. In the era of Big Data, a decision maker may decide to search for an answer within 5% of the optimal solution, for example. Such a solution is reached much faster than searching for a fully optimal solution.

Wrap Up

Many business problems require integer or mixed integer solutions. This chapter discussed various types of IP models: linear, nonlinear, and goal programming. The IP formulation and Solver solution is provided to demonstrate similarities and differences between IP models and their respective original models. In addition, the chapter discusses and offers a general formulation of two common IP models known as the assignment problem and the knapsack problem. The assignment method is used to assign resources to specific tasks assuming that only one resource can be assigned in a given task and only one task can be assigned to a given resource. It requires that decision variables are not only integers, but also binary values. The knapsack problem referred to a hiker deciding to pack the most valuable items to carry considering a weight limit. Several business applications of both the assignment problem and the knapsack problem are also provided.

The formulation of IP models is simple. The requirement that all or some decision variables must be integers is added to any original LP, NLP, or GP model to create an IP model. Setting up the problem with Solver is also relatively simple: A set of integer constraints for all or some decision variables is added in the constraint section. The Solver Parameters dialog box permits the choice of *int* (for integer) and *bin* (for binary) in addition to ≤, ≥, or =.

To strictly enforce integer values for the final solution, you (the decision maker) must also set the tolerance level for integer constraints (under Options) to zero, as shown in Figure 6-11. The solution procedure is longer when the tolerance value is zero or generally tighter. For large problems, setting the tolerance level to 5% is suggested. Once a solution is found, then you can experiment by setting the tolerance level tighter.

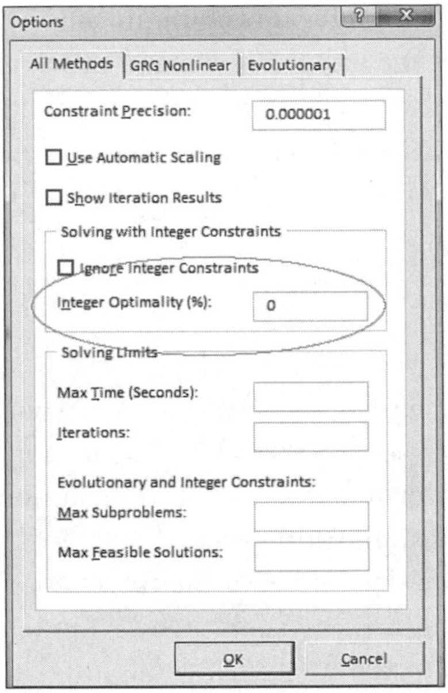

Figure 6-11 Setting the tolerance level for integer constraints

Review Questions

1. Discuss the assumption of divisibility and how it is violated in the case of IP models. Why might it be necessary to require integer values for the solution of some programming models? Provide examples to support your ideas.

2. Describe situations where it is okay for you to simply round the solution results instead of enforcing integer values in the model formulation.

3. Explain the impact of adding integer constraints in the nature of regular programming models. Provide examples to illustrate your answers.

4. Describe different types of IP models and provide situations when a decision maker might choose them. Also, describe situations when mixed IP models are used.

5. Describe the challenges of requiring integer or mixed integer solutions for nonlinear programming models. When can the decision maker select the evolutionary solution method?

6. Discuss the main assumptions of the knapsack problem. Provide business situations that can be modeled with the knapsack problem.

7. Discuss the main assumptions of the assignment problem. Provide business situations that can be modeled with the assignment problem.

8. What should you anticipate when solving integer programming models regarding the values of the objective function and decision variables? Offer a logical explanation for these expectations.

9. The GRG algorithm may not always produce an optimal solution when trying to solve a nonlinear IP model. What can you do to explore alternative solutions and check for possible global optimum?

10. The number of decision variables in the assignment problem tends to grow significantly. Conduct an Internet search and further explore the premium version of Solver or other software programs that can be used to handle large programming models. Summarize the results in a table format.

Practice Problems

1. There are five runners qualified to run the 800 × 4 relay, but only four slots are needed for the race. This means one runner will become an alternate in the event any of the others cannot run the day of the track meet. The following table contains the expected times for each runner based on the data from the season. The track coach is trying to select the runners and the

order of them for the upcoming state meet. Some runners are stronger at the start or finish and others are better in the middle. The assignment program will help him determine which runners to place where. His goal is to have an overall time of 9.05 or less to have a chance at first place.

Runner\Leg	Leg 1	Leg 2	Leg 3	Leg 4
Fowler	2.20	2.18	2.35	2.29
Clark	2.19	2.23	2.25	2.39
Norman	2.22	2.30	2.29	2.42
Houston	2.23	2.36	2.40	2.45
Massey	2.25	2.40	2.50	2.59

a. Formulate the problem as an assignment problem.

b. Create an Excel template and apply Solver to select the runners and their leg assignments in the relay. Do you think the track coach will be able to reach the goal of having an overall time below 9.05?

c. Assume that Fowler will not be able to compete due to an injury during the practice. Rerun the model to assign the best leg assignments for the remaining four runners. Will the coach still be able to reach the overall time goal?

2. You are the operations manager of a health insurance company and are implementing new benefit plans to remain competitive in the new Healthcare Marketplace. An important aspect of the process is to enroll as many customers as possible in the selected plans. The advertising team has been tasked with reaching the maximum total audience via television, newspaper, and radio advertisements without violating budgetary concerns. Television advertisements reach 100,000 customers and cost $9,250; newspaper advertisements reach 80,000 customers and cost $8,000; short 30-second radio advertisements reach 30,000 customers and cost $2,900; and long 1-minute radio advertisements reach 60,000 customers and cost $3,600. The budget for the advertising campaign is $100,000. Based on past experience, the company has limited the number of TV ads to no more than 12, newspaper ads to no more than 5, short 30-second radio ads to no more than 25, and long 1-minute radio ads to no more than 20. Also, no more than $36,000 should be allocated to radio ads; however, at least 5 radio ads will be needed in total for this campaign.

a. Formulate the advertisement problem as a knapsack model where the goal is to maximize the number of customers reached within the total budget for the advertising campaign. Create an Excel template and apply Solver to select the number of advertisements for each media.

b. Modify the previous formulation to also consider the constraints related to the maximum number of advertisement for each media, the maximum budget for radio advertisements, and the minimum number of radio advertisements. Answer the following questions:

- How many customers can be potentially reached via all media advertisements?

- Will the company reach more customers if it decides to increase the total budget from $100,000 to $120,000? Explain why.

3. Consider the animal shelter problem from Chapter 2 (problem 2). There is a new type of food, Pedigree, which you must consider purchasing. While data about Purina and Natural Choice are the same, Pedigree food costs $110 per bag. Each bag of Pedigree weighs 180 pounds and can feed as many as 16 animals per month. You have the same budget of $2,800 for this purchase. The facility has room for no more than 2,000 pounds and can board an average of 100 animals per day during a typical month. How many bags of each type of food should you buy to minimize the purchasing cost?

a. Download the file *ch6_P3animalshelter* and adjust the template to include Pedigree in the model. Solve the problem and determine the optimal number of each type of food bags to be purchased each month.

b. Enforce the integer requirements for each decision variable and rerun the model. Will the solution change? What is the value of the objective function (total purchasing cost) for the previous solution?

c. Assume that the goal of the model will change to maximize the number of animals to be fed during the month. The constraints of the model are space and budget limitations. Rerun the model and determine the optimal number of each type of food bags to be purchased each month. Are resources represented by these two constraints fully utilized? Explain why.

4. Consider the toy manufacturing problem from Chapter 2 (problem 4). The following is a summary of the problem description:

Each production lot of truck toys requires six pounds of plastic, ten hours of labor, and ten hours of machine time. A car toy lot requires eight pounds of plastic, eight hours of labor, and four hours of machine time. There are 72 pounds of plastic available, 80 hours of labor available, and 60 hours of machine time for each day. The profit for either toy is $1,000 per lot.

Assuming that the manufacturer cannot produce fractions of toy lots, adjust the linear programming formulation and solve the problem graphically. Answer the following questions:

a. What is the optimal number of toy truck lots and toy car lots to be purchased each day?

b. What is the value of the objective function (total profit) for the previous solution?

c. Identify binding and not binding constraints for the optimal solution.

5. Consider the same toy manufacturing company, but suppose the manager wants to consider several priorities. This problem was presented in Chapter 5 (problem 5) and the Excel file is *ch6_P5toys* and can be downloaded from the companion website at www.informit.com/title/9780133760354.

a. If you have not solved this problem before, formulate and solve the model as a goal programming model.

b. Add the requirement that the solution values be integers and solve the integer goal programming model.

c. Compare the results from steps a and b and provide managerial recommendations for the operations manager.

6. Consider the nonlinear programming model used in problem 1 from Chapter 4. The Excel file *ch6_P6chair_table* has the template and the final solution. As a reminder, the following is the problem description for the model:

A chair manufacturer wants to determine how many units for each category the company should produce during the next month. Each single-seat chair needs an average of $30 of raw materials and takes an average of eight hours to make. Each multiple-seat chair needs an average of $70 of raw materials and takes an average of ten hours to make. There are 10,000 hours available and a $20,000 budget to purchase raw materials each month. The fixed cost per unit decreases when the production volume increases and the Excel file contains the cutoff points for the profit.

As shown, the optimal solution suggests that fractions of tables and chairs be produced to maximize the profit. Adjust the model to ensure integer values for the number of tables and chairs to be produced.

7. Consider the golf club manufacturing model described in Chapter 4 (problem 6). The following is a brief description of the problem. For more information, you may also download the file *ch6_P7ironsets* from the companion website.

Consider the golf manufacturer problem from the previous chapter. The goal of the operations manager is to determine how many iron sets to produce each month so the company can maximize the revenue. The amount of machine hours is still 20,000 and the company still pays its workers in the iron set assembly line an average of $40 per hour. The minimum production quota of 500 units and the maximum quota of 1,000 units are still required; however, the manager is allowed to offer a 20% discount to those iron sets that sell over 600 units per month.

a. Formulate the problem as an integer nonlinear programming model.

b. Use Solver to generate a solution and analyze the results displayed in the Answer Report.

7

Business Analytics with Shipment Models

Chapter Objectives

- Explain shipment models and their structure as special cases of mathematical programming
- Demonstrate the use of network diagrams for the transportation and transshipment problems
- Learn how to expand transshipment models by adding layers of warehouses or distribution centers
- Learn how to check for the feasibility of the transportation models before even attempting to formulate and solve the model
- Use Excel templates to set up and formulate transportation or transshipment models as linear programming models
- Demonstrate the Solver solutions for the transportation and transshipment problems
- Discuss the benefits of using a network diagram to represent a transportation or transshipment model
- Perform sensitivity analysis and explain the meaning of the shadow price and reduced cost in the solution of transportation models
- Discuss practical recommendation when exploring Big Data with transportation and transshipment models

Prescriptive Analytics in Action: Danaos Saves Time and Money with Shipment Models[1]

Danaos Corporation is a leading international owner of containerships and currently owns more than 54 containerships. In total, these ships carry over 320 thousands standard containers.[2] The company operations are global and include chartering, crewing, and vessel management. Just as in many other shipping companies, planners at Danaos seek to minimize the overall cost of operations and offer the fastest voyages possible. In 2011, the Danaos Corporation set the goal of improving the whole fleet utilization and maximizing its routing revenues.

A sophisticated management program, known as ORISMA (Operations Research in Ship Management), was designed by management consultants and professional experts at Danaos. ORISMA combined internal and external sources and information such as financial data, hydrodynamic models, and weather forecasting. The program included several components, among others, linear and nonlinear programming. The implementation of the program led to optimum solutions at the operational, tactical, and strategic level. Specifically, the company reported an average increase of 8% of net profit, which was approximately equal to $1 million per vessel per year. The anticipated cost savings for the year 2012 for all ORISMA clients was estimated at about $500 million.

Introduction

Decision makers today seek to optimize not only production or service operations, but also those operations that transport goods from plants, to warehouses, to distribution centers, or other destinations. The shipment of such goods includes organizational, intraorganizational, and interorganizational shipments and count for a significant part of the costs of products and services. This chapter discusses two

---

[1] https://www.informs.org/Sites/Getting-Started-With-Analytics/Analytics-Success-Stories/Case-Studies/Danaos-Corporation

[2] http://www.danaos.com/vital_stat.php

types of shipment problems: transportation and transshipment models. The transportation model is based on shipment problems that have two layers: sources, such as machines, plants, and cities; and destinations, such as other machines, warehouses, or cities. Transshipment problems, on the other side, are more complicated. They include more than two layers of shipments. For example, a transshipment problem with four layers may include sources (such as plants), first destinations (warehouses), second destinations (distribution centers), and third destinations (retailers).

Shipment models can be represented with network diagrams that contain nodes and arcs. The nodes usually represent locations, cities, warehouses, or machines, which designate either sources or designations. Sometimes, network nodes may represent points in time. The diagram arcs represent the flow of goods, materials, people, or funds. In addition, both transportation and transshipment models can be formulated as linear programming models.

The Transportation Model

Consider a company that wants to ship products from its warehouses to its retail customers. The logistics manager wants to assign shipment quantities from each warehouse (source) to each retailer (destination) and meet the customer demand at each destination, not exceed available capacity in each source, and minimize transportation cost. This logistic problem can be formulated and solved with the transportation model.

General Formulation of the Transportation Model

Let:

- m = number of sources in the model
- n = number of destinations in the model
- $i = 1, 2 \dots m$, index indicating each individual source
- $j = 1, 2 \dots n$, index indicating each individual destination
- c_{ij} = transportation cost per unit from source i to destination j

- D_j = amount of units demanded by destination j
- S_i = amount of units available at source i

Decision variables for the transportation model can be defined as:

- x_{ij} = amount of units to be transported from source i to destination j during the planning period

The transportation model can be formulated as:

$$\text{Minimize } Z = \sum_{i=1}^{m}\sum_{j=1}^{n} c_{ij} x_{ij} \tag{7.1}$$

subject to:

$$\sum_{i=1}^{m} x_{ij} \geq D_j \text{ for all } j = 1,2,3, ..., n \tag{7.2}$$

$$\sum_{j=1}^{n} x_{ij} \leq S_i \text{ for all } i = 1,2,3, ..., m \tag{7.3}$$

$$x_{ij} \geq 0 \tag{7.4}$$

The transportation model seeks to minimize the value expressed via equation (7.1), which is the scalar product of the quantities shipped from source i to destination j with their respective transportation costs per unit c_{ij}. In specific cases, the contribution coefficients c_{ij} may represent profits, revenues, or other measures that transform the objective function to having a maximization goal. Constraints (7.2) ensure that the sum of units from all sources delivered to destination j is greater than or equal to the demand of destination. Constraints (7.3) ensure that the sum of units to all destinations originated from source i is less than or equal to the available units in that source. Constraints (7.4) represent non-negativity constraints.

As seen in the preceding formulation, the transportation model is a special case of LP models. The most notable characteristic of the transportation model is the value of the technological coefficients. As shown, technological coefficients in the transportation model are all equal to one. This characteristic is used to seek alternative solution methodologies. However, these methodologies fall beyond the practical scope of this book and will not be covered here. As in the previous chapters, the transportation model will be solved via Solver.

Another characteristic of the transportation model is the direction of the inequality constraints. Although the destination constraints (7.2) require the left-hand side value (LHV) to be greater than or equal to the right-hand side value (RHV), the source constraints (7.3) require that the LHV be less than or equal to the RHV. When the total supply for the product generated by all sources is equal to the total demand consumed by all destinations, then both (7.2) and (7.3) can be formulated as equality constraints.

When the total supply for the product generated by all sources is less than the total demand consumed by all destinations, then the model becomes infeasible. In practice, checking for feasibility, ensuring that the total supply of products available for shipping exceeds the total demand, should be the first step before attempting formulation and solution of transportation models. If the problem is infeasible, then the decision maker can create a "dummy" source to equalize the demand and supply. The modeler can assign a very high value to the shipment cost for the units originating from this source. This allows the Solver to first allocate shipments to the real sources and once these sources are exhausted, the shipments are allocated to the "dummy" source. Any value assigned in the "dummy" source row as part of the final solution is considered to be the unmet demand.

Network Diagram of Transportation Models

National Xpress Transportation (NXT) is planning the next-month shipments from its four production plants to its three warehouses. The goal of NXT is to minimize the total shipping cost. Each plant has a monthly production capacity and each warehouse has placed monthly orders that must also be met. The shipping costs per shipment container per each route between plants and warehouses, the monthly production capacity in containers for each plant, and the warehouse demands for the month are shown in Table 7-1.

Table 7-1 Operational Data for the NXT Transportation Model

Plants	Warehouse 1	Warehouse 2	Warehouse 3	Monthly Production Capacity
Atlanta	$60	$35	$65	75
Boston	$55	$30	$70	85
Chicago	$20	$30	$55	95
Denver	$40	$60	$40	80
Monthly Demand	90	120	100	

The previous problem is feasible because the total supply (75 + 85 + 95 + 80 = 335) is greater than total demand (90+ 120 + 100 = 310). As such, there is no need to create artificial (dummy) sources. Figure 7-1 represents the flow diagram for the previous transportation model. On the left side are the four cities where manufacturing plants are located. The numbers below each city indicate the number of product containers that can be produced in the respective city. On the right, the warehouses and their monthly demand for containers is shown. The arrows represent the transportation routes from each plant to each warehouse. The numbers next to each arrow indicate the transportation cost of one container in that specific route.

Solving Transportation Model with Solver

Figure 7-2 shows the Excel template for the previous transportation model. The template has three sections: Parameters, Decisions, and the Objective Function. The Parameters section contains the transportation costs per each route per unit and the total supply at the sources and total demand at the destinations. Sometimes, the modeler has to enforce that shipments cannot be sent from a certain source to a certain destination. In such situations, the modeler assigns a very large number to the shipment cost for that route. Also, the modeler can use the Parameters section to transform an infeasible problem into a feasible one by adding another row as a source. As mentioned earlier, the costs for the cells in that row are significantly large numbers.

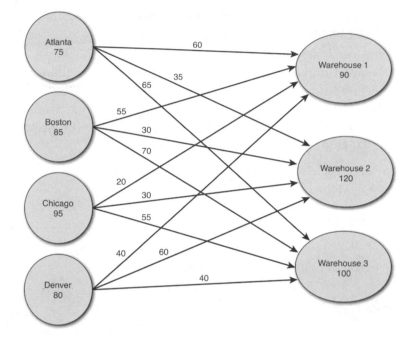

Figure 7-1 Diagram for the NXT transportation model

	A	B	C	D	E	F	G	H
3	Parameters							
4		*From/To*	*Warehouse 1*	*Warehouse 2*	*Warehouse 3*	*Monthly Production Capacity*		
5		Atlanta	60	35	65	75		
6		Boston	55	30	70	85		
7		Chicago	20	30	55	95		
8		Denver	40	60	40	80		
9		*Monthly Demand*	90	120	100			
10								
11	Decisions							
12		*From/To*	*Warehouse 1*	*Warehouse 2*	*Warehouse 3*	*Delivered*		
13		*Atlanta*	0	35	15	50	=SUM(C13:E13)	
14		*Boston*	0	85	0	85	=SUM(C14:E14)	
15		*Chicago*	90	0	5	95	=SUM(C15:E15)	
16		*Denver*	0	0	80	80	=SUM(C16:E16)	
17		*Received*	90	120	100			
18			=SUM(C13:C16)	=SUM(D13:D16)	=SUM(E13:E16)			
19	Objective function							
20			10025	=SUMPRODUCT(C5:E8, C13:E16)				

Figure 7-2 Excel template and initial calculations for the transportation model at NXT

The Decisions section holds the initial values (zeroes) for the decision variables. Row 17 (received) is used to calculate the total units sent to each warehouse as indicated by the formulas shown in the row below. Column F (delivered) is used to calculate the total units originated from each plant as indicated by the formulas shown on the right of the respective cells. The Objective Function section (cell B20) holds the value of Z, which is calculated using the SUM-PRODUCT function, as indicated in the figure.

After the template is created, Solver (see Figure 7-3) is invoked and the parameters are assigned as follows:

- Set Objective: B20
- To: Min
- By Changing Variable Cells: C13: E16

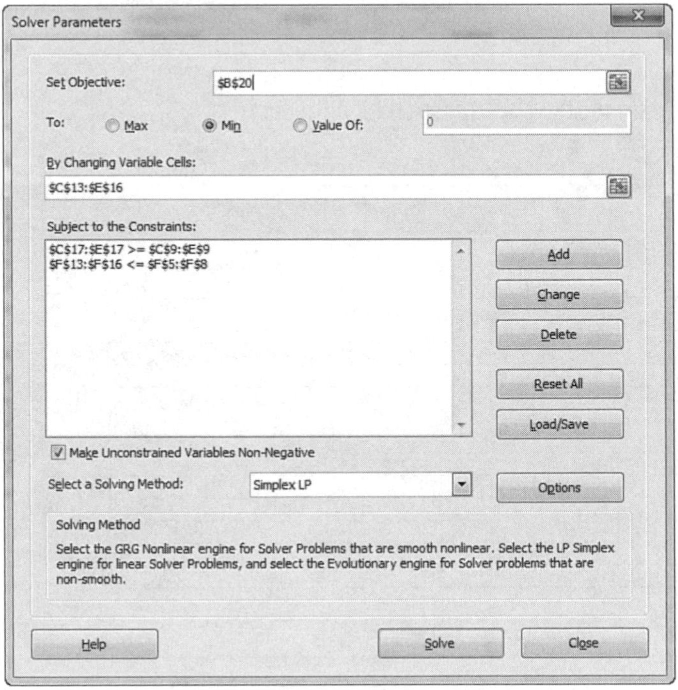

Figure 7-3 Solver Parameters setup for the NXT model

- Subject to the Constraints:
  - C17:E17 >= C9:E9
  - F13:F16 <= F5:F8
- Make Unconstrained Variables Non-Negative: Checked
- Select a Solving Method: Simplex LP

The solution of this model is shown in Figure 7-4, which indicates a minimum transportation cost of $10,025. The solution indicates that the demand for products in each warehouse is completely satisfied. Warehouse 1 will receive 90 containers from Chicago. Warehouse 2 will receive 35 containers from Atlanta and 85 containers from Boston for a total of 120 containers. Warehouse 3 will receive 100 containers from Atlanta (15 units), Chicago (5 units), and Denver (80 units). As shown, the available production capacity in Atlanta (75 containers) is not fully delivered. Only 50 containers are delivered, respectively to Warehouse 2 (35 units) and Warehouse 3 (15 units).

	A	B	C	D	E	F
1	NXC Transportation Model					
2						
3	Parameters					
4		*From/To*	*Warehouse 1*	*Warehouse 2*	*Warehouse 3*	*Monthly Production Capacity*
5		Atlanta	60	35	65	75
6		Boston	55	30	70	85
7		Chicago	20	30	55	95
8		Denver	40	60	40	80
9		*Monthly Demand*	90	120	100	
10						
11	Decisions					
12		*From/To*	*Warehouse 1*	*Warehouse 2*	*Warehouse 3*	*Delivered*
13		*Atlanta*	0	35	15	50
14		*Boston*	0	85	0	85
15		*Chicago*	90	0	5	95
16		*Denver*	0	0	80	80
17		*Received*	90	120	100	
18						
19	Objective function					
20		10025				

Figure 7-4 Final solution for the NXT transportation model

Considering the initial network diagram and ignoring the decision variables that resulted in zero, the modeler can represent the final solution in a simplified network diagram, as shown in Figure 7-5. As shown, the diagram indicates that the requirement to deliver the production capacity in Atlanta is not a binding constraint. All other requirements are binding constraints. Also, the diagram can be used to check that the solution makes sense at each location. For example, at node "Warehouse 2," the total shipment is 120 containers, 35 from Atalanta's plant and 85 from Boston's plant.

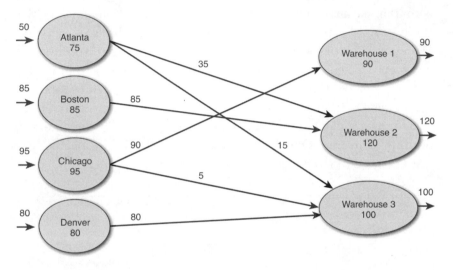

Figure 7-5 Network diagram for the final solution of the transportation model at NXT

Sensitivity Analysis

Just as in LP models discussed in previous chapters, sensitivity analysis can be used to gain more insights about the final solution. The sensitivity analysis for the transportation model is very similar to the one for LP models. This analysis allows the logistic manager to explore the conditions under which the changes in the transportation costs per unit or constraints will not alter the solution.

Changes in the Right-Hand Side Values

The transportation model has two types of constraints: demand and supply constraints. The Sensitivity Report shown in Figure 7-6 provides the shadow prices on each type of constraint. Recall that shadow prices show how the value of the objective function changes when one additional unit of the constraint is acquired. In the case of demand constraints, the shadow price indicates how much it will cost to ship a marginal container to the respective warehouse. For example, suppose that Warehouse 1 demands another two containers. How would that affect the objective function? The shadow price 30 indicates that the value of the objective function will increase by $30 \times 2 = \$60$. However, the marginal cost of $30 remains true when the RHV for this warehouse ranges from 90 - 15 = 75 (allowable decrease of 15) to 90 + 5 = 95 (allowable increase of 5). If the RHV for this warehouse increases, let's say by 10 containers, the logistics manager needs to rerun the model to assess the impact of such a change in the value of the objective function.

22	Constraints						
23			Final	Shadow	Constraint	Allowable	Allowable
24	Cell	Name	Value	Price	R.H. Side	Increase	Decrease
25	$C$17	Received Warehouse 1	90	30	90	5	15
26	$D$17	Received Warehouse 2	120	35	120	25	35
27	$E$17	Received Warehouse 3	100	65	100	25	15
28	$F$13	Atlanta Delivered	50	0	75	1E+30	25
29	$F$14	Boston Delivered	85	-5	85	35	25
30	$F$15	Chicago Delivered	95	-10	95	15	5
31	$F$16	Denver Delivered	80	-25	80	15	25

Figure 7-6 Sensitivity analysis: changes in the constraints

Notice that the shadow prices for the supply constraints are either zero or a negative value. A zero value for the shadow price indicates that the constraint is not a binding constraint. In other words, the reduction of increase in the value of RHV for Atlanta will not impact the objective function. This statement is true when the RHV value, currently 75, stays above 50 units (allowable decrease of 25 units). The other three sources are represented with binding constraints (final value is the same as the RHV). Any change in the RHV will reduce the value of the objective function as follows: Boston by minus $5, Chicago by minus $10, and Denver by minus $25. Obviously, increasing production capacity in Denver and making more containers available for shipping in Denver is the most attractive option to consider. Again, these shadow prices remain true in the ranges defined by the allowable increase and allowable decrease values.

Changes in the Contribution Coefficients

Solver also produces a Sensitivity Report for decision variables, as shown in Figure 7-7.

6	Variable Cells						
7			Final	Reduced	Objective	Allowable	Allowable
8	Cell	Name	Value	Cost	Coefficient	Increase	Decrease
9	$C$13	Atlanta Warehouse 1	0	30	60	1E+30	30
10	$D$13	Atlanta Warehouse 2	35	0	35	5	5
11	$E$13	Atlanta Warehouse 3	15	0	65	10	5
12	$C$14	Boston Warehouse 1	0	30	55	1E+30	30
13	$D$14	Boston Warehouse 2	85	0	30	5	1E+30
14	$E$14	Boston Warehouse 3	0	10	70	1E+30	10
15	$C$15	Chicago Warehouse 1	90	0	20	30	30
16	$D$15	Chicago Warehouse 2	0	5	30	1E+30	5
17	$E$15	Chicago Warehouse 3	5	0	55	5	30
18	$C$16	Denver Warehouse 1	0	35	40	1E+30	35
19	$D$16	Denver Warehouse 2	0	50	60	1E+30	50
20	$E$16	Denver Warehouse 3	80	0	40	25	1E+30

Figure 7-7 Sensitivity analysis: changes in the decision variables

Recall that the reduced cost for a decision variable is the shadow price of the constraint for that variable. In the transportation model, each variable is subject to non-negativity constraints. As such, the reduced costs indicate how much the cost per unit should change so containers can be assigned to routes that don't have any assignments or cannot be assigned to routes that currently have assignments. For example, currently there is no container moved from Atlanta to Warehouse 1. The current contribution coefficient (transportation cost per unit from Atlanta to Warehouse 1) is $60. Because the reduced cost is 30, that means that the transportation cost must be reduced to $30 (60 - 30) or more so this route can become attractive for shipping assignments. Similarly, the current solution indicates that 35 units must be assigned from Atlanta to Warehouse 2. The current contribution coefficient (transportation cost per unit from Atlanta to Warehouse 2) is $35. If the cost increases to more than $40 (current value plus allowable increase 5), this route might become unattractive for shipping assignments.

The Transshipment Method

Consider a company that wants to ship products from its three plants to its retail stores. The finished products are first distributed from three plants to the company's five warehouses. The warehouse logistics manager stores the products according to potential retail destinations. Upon demand, the finished products are shipped to each of the six retail stores. The company can produce a total of 470 units per month (120 from plant 1, 200 from plant 2, and 150 from plant 3). The total monthly demand from the retail stores is 400 units (respectively 80, 40, 10, 20, 100, and 150 from each retail store). The transshipment problem is graphically represented with the network diagram shown in Figure 7-8.

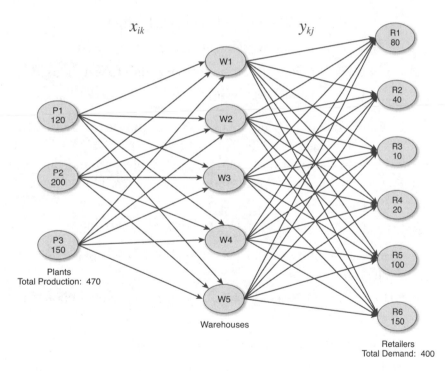

Figure 7-8 Network diagram for the transshipment problem

General Formulation of the Transshipment Model

Let:

m = number of plants in the model (for the sample problem $m = 3$)

l = number of warehouses in the model (for the sample problem $l = 5$)

n = number of retail stores in the model (for the sample problem $n = 6$)

$i = 1, 2 \ldots m$, index indicating each individual plant

$k = 1, 2 \ldots l$, index indicating each individual warehouse

$j = 1, 2 \ldots n$, index indicating each individual store

D_j = amount of units demanded by retail store j

S_i = amount of units available at plant i

c_{ik} = transportation cost per unit from plant i to warehouse k

c_{kj} = transportation cost per unit from warehouse k to store j

Transportation costs from the plants to warehouses (c_{ik}) and from warehouses to retail stores (c_{kj}) are shown respectively in Tables 7-2 and 7-3. In the rightmost column, these tables also indicate the monthly supply from each plant and monthly demand by each retail store.

The logistics manager needs to find two set of variables:

- How many units should be shipped from each plant to each warehouse?
- How many units should be shipped from each warehouse to each store?

As a result, the decision variables for the transshipment model can be defined as:

- x_{ik} = amount of units to be shipped from plant i to warehouse k during the planning period
- y_{kj} = amount of units to be shipped from warehouse k to retail store j during the planning period

Table 7-2 Shipment Costs per Unit from Plants to Warehouses

From/To	Warehouse 1	Warehouse 2	Warehouse 3	Warehouse 4	Warehouse 5	Monthly Production
Plant 1	60	35	65	40	40	120
Plant 2	55	30	70	120	50	200
Plant 3	20	30	55	60	20	150
Total						470

Table 7-3 Shipment Costs per Unit from Warehouses to Retail Stores

To/From	Warehouse 1	Warehouse 2	Warehouse 3	Warehouse 4	Warehouse 5	Monthly Demand
Retail store 1	30	70	70	10	10	80
Retail store 2	20	80	10	60	30	40
Retail store 3	80	40	30	100	30	10
Retail store 4	90	30	20	20	90	20
Retail store 5	10	40	70	60	50	100
Retail store 6	100	20	90	30	60	150
Total						400

The transportation model can be formulated as:

$$\text{Minimize } Z = \sum_{i=1}^{m}\sum_{k=1}^{l} c_{ik} x_{ik} + \sum_{k=1}^{l}\sum_{j=1}^{n} c_{kj} y_{kj} \tag{7.5}$$

subject to:

$$\sum_{k=1}^{l} y_{kj} \geq D_j \qquad \text{for all } j = 1,2,3,\ldots,n \tag{7.6}$$

$$\sum_{k=1}^{l} x_{ik} \leq S_i \qquad \text{for all } i = 1,2,3,\ldots,m \tag{7.7}$$

$$\sum_{i=1}^{m}\sum_{k=1}^{l} x_{ik} - \sum_{k=1}^{l}\sum_{j=1}^{n} y_{kj} = 0 \qquad \text{for all } k = 1,2,3,\ldots,l \tag{7.8}$$

$$x_{ik} \geq 0 \text{ and } y_{kj} \geq 0 \tag{7.9}$$

The transshipment model seeks to minimize total transportation cost from plants to warehouses and from warehouses to retail stores, as shown in (7.5). The Z value is the sum of two scalar products: quantities shipped from plant i to warehouse k with their respective transportation costs per unit (c_{ik}) and quantities shipped from warehouse k to retail store j with their respective transportation costs per unit (c_{kj}). Constraints (7.6) ensure that the sum of units from all warehouses delivered to stores is greater than or equal to the demands in each destination j. Constraints (7.7) ensure that the sum of units to all warehouses originated from each plant i is less than or equal to the available units in that plant. Constraints (7.8) ensure that the amount of units entering each warehouse is equal to the amount of units leaving the warehouse. Finally, constraints (7.9) represent non-negativity constraints for both sets of decision variables.

As seen in this formulation, the transshipment model is an extension of the transportation model. The transshipment model includes one additional layer in supply chain of goods. The model can be adjusted to include more layers as needed for a realistic representation of supply chain. For example, the supply chain of goods from raw materials, to components, to assembly, to warehouse, to distribution centers, to retailers can be formulated as an extended transshipment model with several layers and several sets of decision variables.

Solving the Transshipment Model with Solver

Figure 7-9 shows the Excel template for the transshipment model. Similar to the transportation model, the transshipment template also has three sections: Parameters, Decisions, and the Objective Function. The Parameters section of the transshipment model shown on the left side of the template contains the transportation costs per unit per each route connecting plants and warehouses (cells C5:G7) and transportation costs per unit per each route connecting warehouses and retail stores (cells C11:G16). The template also contains monthly production (available units) from each plant (cells H5:H7) and monthly demand from each retail store (cells H11:H16).

The two sets of decision variables (x_{ik} and y_{kj}) are located on the right side of the template in the shaded area of Figure 7-9. The shaded area in the figure holds the initial values of such decision variables. The sum of units delivered from each plant to all warehouses is calculated in cells P5:P7 and the sum of units delivered to each store from all warehouses is calculated in cells P11:P16. The value of the objective function is placed in cell J20 and is calculated as the sum of two scalar products using the following formula:

$$J20 = \text{SUMPRODUCT (C5:G7, K5:O7)} +$$
$$\text{SUMPRODUCT (C11:G16, K11:O16)} \qquad (7.10)$$

The Solver Parameters setup for the transshipment model is shown in Figure 7-10.

Note the differences between the transportation and transshipment models. The objective function cell in the transportation model (B20) contained the value of one scalar product. The objective function in the transshipment model (cell J20) has two scalar products, as shown in the formula (7.10). In both cases, the decision maker seeks to minimize the total shipment cost.

Parameters: from Plants to Warehouses

From/To	Warehouse 1	Warehouse 2	Warehouse 3	Warehouse 4	Warehouse 5	Monthly Production
Plant 1	60	35	65	40	40	120
Plant 2	55	30	70	120	50	200
Plant 3	20	30	55	60	20	150
Total						470

Parameters: from Warehouses to Retail Stores

To/From	Warehouse 1	Warehouse 2	Warehouse 3	Warehouse 4	Warehouse 5	Monthly Demand
Retail store 1	30	70	70	10	10	80
Retail store 2	20	80	10	60	30	40
Retail store 3	80	40	30	100	30	10
Retail store 4	90	30	20	20	90	20
Retail store 5	10	40	70	60	50	100
Retail store 6	100	20	90	30	60	150
Total						400

Decision variables: X_{ik}

From/To	Warehouse 1	Warehouse 2	Warehouse 3	Warehouse 4	Warehouse 5	Monthly Production	Unused Capacity
Plant 1	0	0	0	100	0	100	20
Plant 2	0	150	0	0	0	150	50
Plant 3	140	0	0	0	10	150	0
Total	140	150	0	100	10		

Decision Variables: Y_{kj}

To/From	Warehouse 1	Warehouse 2	Warehouse 3	Warehouse 4	Warehouse 5	Monthly Demand	Unmet Demand
Retail store 1	0	0	0	80	0	80	0
Retail store 2	40	0	0	0	0	40	0
Retail store 3	0	0	0	0	10	10	0
Retail store 4	0	0	0	20	0	20	0
Retail store 5	100	0	0	0	0	100	0
Retail store 6	0	150	0	0	0	150	0
Total	140	150	0	100	10		

Objective function

17800 =SUMPRODUCT(C5:G7, K5:O7)+SUMPRODUCT(C11:G16, K11:O16)

Figure 7-9 Excel template and initial solution for the transshipment model

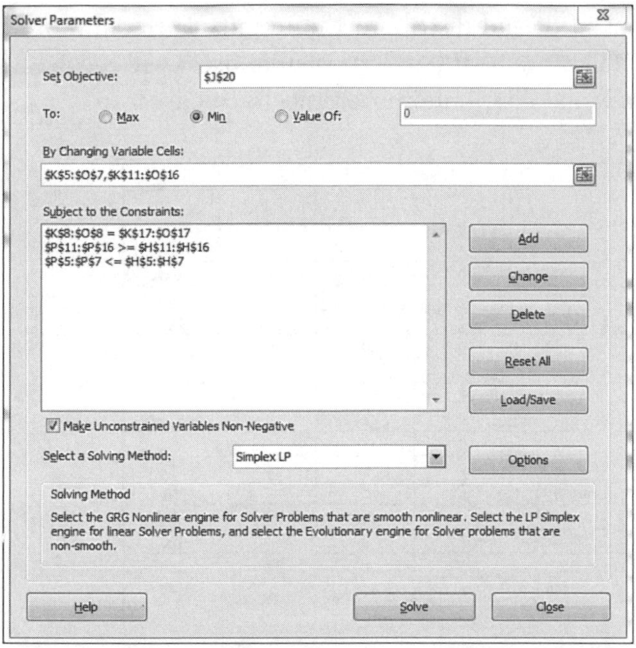

Figure 7-10 Solver Parameters setup for the transshipment model

In the By Changing Variable Cells field, the transportation model has one set of decision variables, whereas the transshipment model has two sets of variables: K5:O7 and K11:O16. Like in the transportation model, there are two set of constraints, which ensure that (a) the shipped units do not exceed the available units produced by plants and (b) the shipped units from warehouses meet or exceed store demands. Unlike in the transportation model, there is a third constraint in the transshipment model. Constraint K8:O8 = K17:O17 ensures that for each warehouse, the amount of units received from plants is equal to the units delivered to retail stores.

The optimal solution of the transshipment model is shown in Figure 7-11. The overall shipment cost is $17,800. The demand for units in each store is completely satisfied. Plant 1, for example, delivers 100 units to Warehouse 1. Note that production capacity for Plant 1 is 120 units, so not all available units are used and the constraint for Plant 1 is not a binding constraint. Those 100 units that Warehouse 1 receives will be distributed to retail store 1 (80 units) and retail store 2 (20 units). Warehouse 2 will receive 150 units (from plant 2) and distribute those units to Retail Store 6. Warehouse 1 will receive

140 units (from plant 3) and distribute those units to Retail Store 2 (40 units) and Retail Store 5 (100 units). Warehouse 5 will receive 10 units (from plant 3) and distribute those units to Retail Store 3.

Decision variables:		Xik				
From/To	Warehouse 1	Warehouse 2	Warehouse 3	Warehouse 4	Warehouse 5	Monthly Production
Plant 1	0	0	0	100	0	100
Plant 2	0	150	0	0	0	150
Plant 3	140	0	0	0	10	150
Total	140	150	0	100	10	
Decision Variables:		Ykj				
To/From	Warehouse 1	Warehouse 2	Warehouse 3	Warehouse 4	Warehouse 5	Monthly Demand
Retail store 1	0	0	0	80	0	80
Retail store 2	40	0	0	0	0	40
Retail store 3	0	0	0	0	10	10
Retail store 4	0	0	0	20	0	20
Retail store 5	100	0	0	0	0	100
Retail store 6	0	150	0	0	0	150
Total	140	150	0	100	10	
Objective function						
17800	=SUMPRODUCT(C5:G7, K5:O7)+SUMPRODUCT(C11:G16, K11:O16)					

Figure 7-11 Final solution for the transshipment model

The final solution can also be presented in a simplified network diagram (see Figure 7-12), which ignores the routes where the final value of the decision variables is zero.

Exploring Big Data with Shipment Models

In a recent survey of logistics organizations, 60% of the respondents stated that they were planning to invest in Big Data analytics within the next five years.[55] The availability of Big Data in the transportation industry offers challenges and opportunities for data scientists. Using Radio Frequency Identification (RFID) technology, the decision maker has instant access to delivery times, resource utilizations, geographical coverages, and delivery statuses. It is projected that RFID technology will continue to have the largest single growth in the years to come followed by optimization software technologies.[55]

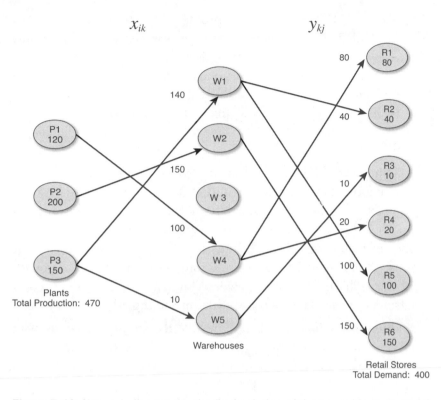

Figure 7-12 Network diagram for the final solution of the transshipment model

The implementation of advanced technologies in transportation and transshipment models allows decision makers to retrieve massive real-time information, which originates from sensors, external databases, and mobile devices and combine them to operate delivery resources at maximum levels of efficiency. The execution of optimization procedures can reroute vehicles on the go, and, as a result, drivers can receive instant direction updates in their onboard navigation systems to the next best destination.[56]

Wrap Up

This chapter discussed two types of shipment problems: transportation and transshipment models. These two models can be used to minimize shipment costs of goods within an organization or between organizations. The transportation problem includes models that have shipment of goods between two layers (sources and destinations). Transshipment problems also include sources and destinations, but also include intermediate layers, such as warehouses and distribution centers. Shipment models can significantly lower the cost of moving goods from the destinations to the sources and, as a result, are an important tool for the logistics manager to increase the efficiency of supply chains.

Shipment models can be represented with network diagrams and, consequently, their formulation and solution interpretation becomes very practical and intuitive. In addition, transportation and transshipment models can be formulated as a special case of LP models. The values of technological coefficients and the constraint formulations are used to develop alternative solution methodologies for shipment problems. This chapter uses Solver to build a template and find optimal solutions for both the transportation and transshipment models.

Often, shipment models require that the amount of goods available generated by sources are greater than or equal to the demand for these goods. However, sometimes, the opposite requirement is true: The logistics must assign shipments that minimize overall costs while utilizing all products available and not exceeding market demand. Also, specific situations may require that the amount of goods available generated by sources is exactly the same as the demand for these goods. Establishing the correct direction for the constraints allows the modeler to formulate both cases. In addition, the modeler is able to check for the feasibility of the problem before attempting to formulate or solve the model. The logistics manager can also ensure that there are no shipments from a certain source to a certain destination by assigning a very large value to the cost of that specific route. The LP simplex method will not allocate units to the route while attempting to minimize the overall cost.

The Solver methodology is similar to the one covered in the previous linear programming chapters. Once a solution is found, the model can visually represent the optimal allocations via a simplified network diagram. Also, sensitivity analysis is similar to the sensitivity analysis discussed in the LP models. The Sensitivity Report can be used to identify the range for the right-hand side within which a precise impact in the objective function can be calculated. Also, the report can be used to identify the range for the contribution coefficients for which the optimal values of the decision variables still remain unchanged.

The transshipment model searches for two sets of decision variables: number of units to be shipped from each source to each intermediary destination (warehouse, distribution center) and number of units to be shipped from each intermediary destination to the final destination. Often, the transshipment models include more than three layers, but the formulation concept remains the same: an equality constraint that balances the number of units entering an intermediary destination with the number of units leaving the same location. Also, the value of the objective function is the sum of two or more scalar products of the costs of shipment with their respective decision variables.

Review Questions

1. Discuss the structure of transportation and transshipment models. How does the transshipment model differ from the transportation model?

2. Transportation and transshipment models are a special case of LP models. What features make shipment models a special case of linear programming? Can transportation or transshipment models be formulated as nonlinear programming models?

3. Expand the transshipment model discussed in the chapter by adding another layer in the model. Assign your own locations and respective shipment costs to and from those locations. Formulate the new problem as an LP model.

4. Can transportation or transshipment models be formulated as linear programming models with a maximization objective function? Provide examples to support your answer.

5. How can the modeler check for the feasibility of the transportation model before even attempting to formulate and solve the model?

6. What is a "dummy source" or a "dummy destination" when attempting to solve a transportation model? What cost values would you suggest in the "dummy cells"? Why?

7. Discuss the benefits of using a network diagram to represent a transportation or transshipment model.

8. Discuss the differences in the general Excel templates used to set up transportation and transshipment models. Consider the differences in the set of decision variables, constraints (in Solver Parameters), and objective functions.

9. What is the meaning of the shadow price in the sensitivity analysis of the transportation model? Is there a difference in the shadow prices between demand constraints and supply constraints? How can the logistic manager use the shadow price to analyze the solution results?

10. What is the meaning of the reduced cost for decision variables in the sensitivity analysis of the transportation model? Is there a difference in the reduced costs between transportation models and traditional LP models? How can the logistic manager use the reduced costs to analyze the solution results?

Practice Problems

1. You are the logistics manager of a shoe manufacturer. The company has three production plants in Dallas, Charlotte, and San Diego, which ship daily to three warehouses in Atlanta, Boston, and Denver. The overall goal is to minimize daily cost of shipping from plants to warehouses. The following table contains the supply for each production plant, demand for each warehouse, and shipping cost per box per each route between plants and warehouses:

From/To:	Atlanta	Boston	Denver	Supply
Dallas	$2.30	$1.75	$4.50	**3,000**
Charlotte	$2.90	$3.35	$3.05	**3,500**
San Diego	$1.99	$2.80	$3.65	**5,000**
Demand	**5,000**	**4,000**	**2,500**	

    a. Represent the problem with a network diagram.

    b. Use Excel to create a template and calculate total transportation costs.

    c. Use Solver to suggest how many boxes should be shipped from each plant to each warehouse.

    d. Perform sensitivity analysis and provide managerial recommendations for future shipments.

    e. Represent the solution with a simplified network diagram.

    f. Assume that the demand in Denver increases from 2,500 to 3,000 boxes. How will this change impact the distribution?

2. You are the logistics manager of a brewing company that has locations in several cities. Each location brews beer and serves as a retail restaurant. Some locations, however, do not have the capacity to supply their own restaurant demands. Some other locations can produce more than their restaurants can consume. Your goal is to ship barrels of beer from sources to restaurant destinations via the shortest possible route as necessary to meet the daily demand. The daily demand and distance between cities is shown in the following table:

From/To	Desert Ridge, AZ	Charlotte, NC	Chicago, IL	La Jolla, CA	Denver, CO	Long Beach, CA	Des Moines, IA	Daily Production Capacity
Desert Ridge, AZ	0	2,452	2,051	470	976	564	1,672	4.5
Charlotte, NC	2,452	0	906	2,870	1,874	2,918	1,257	3
Chicago, IL	2,051	906	0	2,493	1,203	2,434	414	4.5
La Jolla, CA	470	2,870	2,493	0	1,294	135	2,094	3
Denver, CO	976	1,874	1,203	1,294	0	1,235	806	3
Long Beach, CA	564	2,918	2,434	135	1,235	0	2,036	6
Des Moines, IA	1,672	1,257	414	2,094	806	2,036	0	4.5
Daily Demand	1.5	3	6	4.5	4.5	4.5	3	

a. Prepare a network diagram to represent the problem graphically.

b. Prepare an Excel template and calculate total transportation costs for an initial solution.

c. Use Solver to generate an optimal solution.

d. Perform sensitivity analysis to offer managerial recommendations and possible what-if scenarios.

e. Assume that only full barrels of beer can be shipped between cities. Add the *integer* requirements to the model and observe the impact of this change in the solution process.

3. A propane gas distributor has two sources of supply located in Chattanooga and Nashville. The distributor ships its propane tanks from four warehouse locations: Dalton, Nashville, Cleveland, and Omaha. Currently, the company is exploring several distribution centers to ship propane tanks closer to individual retailers. The distribution centers are located in San Diego, Houston, Atlanta, Newark, and Chicago. The distributor wants to find out how many propane tanks should be shipped from each source to each warehouse and then from each warehouse to each distribution center every week so the overall shipping cost is minimized.

The costs to ship one tank from the sources to the warehouses and the supply levels are shown in the following table:

From/To	Dalton	Nashville	Cleveland	Omaha	Weekly Supply
Chattanooga	$5.00	$7.00	$9.50	$9.00	350
Nashville	$6.00	$0.00	$10.50	$11.50	450

The costs to ship one tank from the warehouse to the distribution centers and the demand levels are shown in the following table:

From/To	San Diego	Houston	Atlanta	Newark	Chicago
Dalton	$9.50	$7.50	$3.50	$6.00	$6.50
Nashville	$10.00	$6.50	$4.50	$8.50	$8.00
Cleveland	$4.00	$6.00	$9.50	$10.00	$8.50
Omaha	$12.50	$10.00	$9.50	$3.50	$4.50
Weekly Demand	100	100	200	200	150

    a. Formulate the transshipment model as a linear programming model.

    b. Represent the problem with a network diagram.

    c. Use Excel to create a template and calculate total transportation costs.

    d. Solve the problem and perform sensitivity analysis to generate managerial recommendations.

    e. Represent the solution with a simplified network diagram.

    f. Assume that the weekly demand in San Diego and Houston doubles from 100 units to 200 units each. Will this change impact the solution? Rerun the model and interpret the results.

4. An Energy Production Company (EPC) has ten fossil plants that must receive a certain amount of coal per week to sustain energy generation. Each plant, respective region, and weekly demand is as follows:

Plant	Region	Demand
A	East	450 tons/week
B	West	100 tons/week
C	West	50 tons/week
D	East	135 tons/week
E	West	75 tons/week
F	East	125 tons/week
G	East	150 tons/week
H	East	900 tons/week
I	South	50 tons/week
J	South	50 tons/week

EPC has three main coal suppliers, all of which have different capacities in regard to how many tons of coal they can deliver per week. EPC pays an average of $10,000 per ton with each supplier. The supplier capacities are as follows:

SCSX	1,650 tons/week
SUSX	800 tons/week
SRTV	1,225 tons/week

The coal must first be brought to a regional storage facility for operational reasons, such as quality inspection and inventory control. However, each regional warehouse has a maximum processing capacity as follows:

EAST	900 tons/week
WEST	1,500 tons/week
SOUTH	1,350 tons/week

Generally, coal is transported to plants from their respective regional storage facility. However, there are times when demand exceeds the processing capacity at the regional storage facility, and coal must be transported from another regional storage facility. There is an additional cost associated with doing this. The costs are as follows:

Region	Percent Surcharge
EAST-WEST	33% surcharge
EAST-SOUTH	25% surcharge
WEST-SOUTH	15% surcharge

How should EPC plan its transportation of coal between the suppliers, regional storage facilities, and coal plants to minimize the overall costs, under the limitations of regional storage facilities and supplier capacities while still meeting demand at each coal plant? Specifically:

    a. Formulate the transshipment model as a linear programming model.

    b. Represent the problem with a network diagram.

    c. Use Excel to create a template and calculate total transportation costs.

    d. Solve the problem and perform sensitivity analysis to generate managerial recommendations.

    e. Represent the solution with a simplified network diagram.

5. A carpet manufacturer produces two types of carpets (Nylon and Polyester) and has three production plants. For any given week, each of the plants can produce the following quantities of each product:

City	Plant #	Nylon	Polyester
Calhoun	1	1,000	2,000
Tifton	2	2,000	3,000
El Paso	3	1,000	1,000

The manufacturer has distribution centers strategically located in eight states: Florida (FL), Tennessee (TN), Texas (TX), California (CA), Minnesota (MN), Wyoming (WY), Indiana (IN), and New Jersey (NJ). The following table shows the weekly demand (in rolls) for Nylon and Polyester for each distribution center:

Distribution Center	Nylon	Polyester
Florida (FL)	600	800
Tennessee (TN)	300	415
Texas (TX)	350	480
California (CA)	762	870
Minnesota (MN)	300	625
Wyoming (WY)	250	410
Indiana (IN)	325	795
New Jersey (NJ)	1,040	1,229

Each type of carpet has different shipping costs per roll due to the weight of each product and limitations on the allowable load per truck/shipment. Shipping costs per Nylon roll from each of the production plants to the various distribution centers are summarized in the following table:

From / To	Florida (FL)	Tennessee (TN)	Texas (TX)	California (CA)
Plant 1	$6.00	$5.00	$6.00	$6.00
Plant 2	$5.00	$6.00	$8.00	$8.00
Plant 3	$8.00	$8.00	$5.00	$5.00

From / To	Minnesota (MN)	Wyoming (WY)	Indiana (IN)	New Jersey (NJ)
Plant 1	$5.00	$6.00	$5.00	$5.00
Plant 2	$6.00	$8.00	$6.00	$6.00
Plant 3	$8.00	$5.00	$8.00	$8.00

Shipping costs per Polyester roll from each of the production plants to the various distribution centers are summarized in the following table:

From / To	Florida (FL)	Tennessee (TN)	Texas (TX)	California (CA)
Plant 1	$4.00	$3.00	$4.00	$4.00
Plant 2	$3.00	$4.00	$6.00	$6.00
Plant 3	$6.00	$6.00	$3.00	$3.00

From / To	Minnesota (MN)	Wyoming (WY)	Indiana (IN)	New Jersey (NJ)
Plant 1	$3.00	$4.00	$3.00	$3.00
Plant 2	$4.00	$6.00	$4.00	$4.00
Plant 3	$6.00	$3.00	$6.00	$6.00

The manufacturer wants to determine how many rolls of each product should be shipped from each manufacturer to each distribution center. The optimal solution must minimize the overall shipment cost and meet the weekly demand of the distribution centers. Specifically:

a. Formulate the transshipment model as a linear programming model. Note that the decision variables must have three indexes: carpet type (i), plant (j), and distribution center (k).

    b. Represent the problem with a network diagram. Create one diagram for each product type.

    c. Use Excel to create a template and calculate total transportation costs.

    d. Solve the problem and perform sensitivity analysis to generate managerial recommendations.

    e. Represent the solution with two simplified network diagrams.

6. Hydro Electric Power (HEP) is a local power distributor for the Southeast region of the United States. The company has four hydro dams that are generating power for five power distributors (PD). The following table shows the maximum power that can be generated by each plant during the week:

Dam	Maximum Power Capability
Upper River (Dam 1)	135 (millions kwh)
Tennessee River (Dam 2)	150 (millions kwh)
Upper Creek Fall (Dam 3)	140 (millions kwh)
Bend Stream (Dam 4)	160 (millions kwh)

The demand by the power distributors is summarized in the following table:

Power Distributors	Weekly Demand for Power
PD1	145 (millions kwh)
PD2	120 (millions kwh)
PD3	130 (millions kwh)
PD4	130 (millions kwh)
PD5	150 (millions kwh)

The distance between dams and power distributors is shown in the following table:

	Dam 1	Dam 2	Dam 3	Dam 4
PD1	30	70	70	10
PD2	20	80	10	60
PD3	80	40	30	100
PD4	90	30	20	20
PD5	10	40	70	60

a. Formulate the problem as a linear programming model. The goal is to minimize the overall transportation distance of electric power from dams to power distributors.

b. Represent the problem with a network diagram. Is this problem feasible?

c. Use Excel to create a template and calculate total distribution distance.

d. Solve the problem and perform sensitivity analysis to generate managerial recommendations.

e. Represent the solution with a simplified network diagram.

8

Marketing Analytics with Linear Programming

Chapter Objectives

- Understand the role of marketing analytics as part of business analytics
- Explain the recency, frequency, and monetary value approach model as a descriptive marketing analytics tool
- Demonstrate how to use Excel to classify customers into recency, frequency, and monetary value clusters
- Apply linear programming models to determine segments of customers that must be reached to maximize the profits under budget constraints
- Discuss the challenges of implementing marketing analytics in the era of Big Data

Prescriptive Analytics in Action: Hewlett-Packard Increases Profit with Marketing Optimization Models[1]

In 2005, Hewlett-Packard (HP) established HPDirect.com with the goal of utilizing the Internet to increase sales. Building such capability proved to be a challenge for HP because the existing systems,

[1] https://www.informs.org/Sites/Getting-Started-With-Analytics/
Analytics-Success-Stories/Case-Studies/Hewlett-Packard3

processes, and people were tailored toward large organizations and retailers, and not necessarily toward individual customers. HPDirect. com needed to increase its volume of online sales, conversion of visits to transactions, return visits, and order size. Ultimately, these goals can translate to more frequent purchases, more recent transactions, and more money spent by customers in each transaction.

Data scientists at HP Global Analytics used mathematical programming and other optimization techniques to quantify the "impact that online marketing activities have on customers" and "prioritize marketing dollars across online marketing activities." The proposed models helped improve the average conversion rate from 1.5% to 2.5% and increased the order size by 20%.

Introduction

In the previous chapters, mathematical programming models have dealt mostly with manufacturing or logistics activities in the organization. This chapter introduces the use of linear programming models for marketing purposes. Specifically, the chapter demonstrates how LP models can be used to augment the analysis of data generated by customer transactions from predictions to optimizations.

As shown in Figure 8-1, marketing analytics is an important part of business analytics and can be used to make better business decisions using real market feedback. Automatic capturing of online and offline customer transactions has led to an explosion of customer data. This data can be used to not only generate valuable information about customers, but also information about their buying patterns and customer behavior. This wealth of information can be used by marketing analysts to design better marketing campaigns and increase their return on investment. It is estimated that marketing analytics will increase 60% in 2015 compared with the year before.[57]

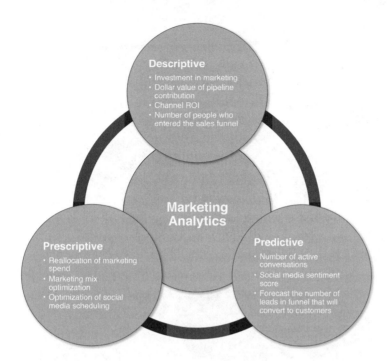

Figure 8-1 The domain of marketing analytics

Figure 8-2 provides a closer view of marketing analytics. This figure depicts three dimensions of marketing analytics as recently discussed by Mawhinney:[57] descriptive, predictive, and prescriptive. The descriptive marketing analytics can be used to analyze the effectiveness and efficiency of investments in marketing, marketing contribution to the sales pipeline, return on investment for different marketing channels, and the number of lead, prospects, or actual potential customers. Descriptive marketing analytics is important in marketing campaigns because it is used to help firms identify and improve *current* response rates, conversion rates, and campaign profitability.

Predictive marketing analytics is also very important for marketing campaigns. It can be used to predict *future* response rates, conversion rates, and campaign profitability. For example, marketing predictive analytics can be used to monitor and identify active conversations in social media, assign a sentiment score to each social media site, and predict the conversion rate of lead customers to buying customers.

Figure 8-2 The domain of business analytics

The focus of this chapter is on prescriptive marketing analytics. It covers a series of important decision-making tools for marketing managers. These tools can be used to reallocate future funds for marketing campaigns, provide the best possible mix of marketing channels, and optimize social media scheduling by identifying the best times of the month or day to post an update, when to make a new offer, or when to upgrade to a new account level.

The recency-frequency-monetary value (RFM) framework is an analytical tool that can be used in combination with descriptive, predictive, and prescriptive analytics. According to the RFM approach, for each customer, the date of most recent purchase (R), frequency (F) or number of purchases during a given time period, and average monetary value (M) or amount of purchase are captured and stored. Using descriptive analytics, the marketing analyst can categorize customers into homogenous segments based on their previous purchasing behavior and then design highly customized promotional campaigns for each segment. For example, if a given customer segment shows relatively high values for frequency and monetary value, but a low value for recency, then these customers can be approached with a "we want you back" marketing strategy. For a customer segment with a low monetary value and high values for frequency and recency, an "up-selling" marketing strategy could be applied to generate additional sales revenue.

Visitors to a company's webpage, subscribers to a monthly newsletter, or the number of followers on a Twitter account are helpful metrics to see trends and make predictions. However, a company must focus on actionable metrics, such as who is coming to your site, what those people are doing after they get there, who is converting, what conversions are driving revenue, who is buying multiple times, what is the lifetime customer value, or what are the churn rates. "Companies must measure metrics that matter," says Neil Patel, founder of KISS-Metrics, as quoted by DeMers [58].

These data are important input parameters for the proposed mathematical programming models offered in this chapter. The results of the RFM approach can be used in conjunction with mathematical programming models as an effective prescriptive marketing analytics tools. The objective function for marketing models is the net present value of cash flow expected during a customer's tenure with a firm. This metric is known as customer lifetime value (CLV) and is a valuable marketing measure. CLV also serves as the basis of customer relationship management (CRM) decisions. For example, CLV might be used to determine whether a service policy exception is made for a key account or whether a credit card customer's credit limit or interest rates are increased. Increasing the CLV is a guiding principle for a firm when trying to acquire a new customer or retain existing ones.

CLV is generally higher when a customer purchases frequently, has purchased recently, and the average amount of money spent in purchases is higher. Fader, Hardie, and Lee [59] link the RFM approach with the customer lifetime value. They offer the concept of *iso-value* curves, which groups individual customers across different purchasing history (read different values for R, F, or M) into groups with similar future valuations. Optimization models can be used to maximize future valuation of marketing campaigns under limited campaign resources.

This chapter offers a brief introduction of RFM analysis. Then, a series of linear programming models that seek to increase CLV within marketing budget constraints is introduced. The proposed models combine data provided by RFM analysis and budgeting data and help direct marketers determine whether to continue or curtail their relationship with a given value of recency, frequency, or monetary values.

RFM Overview

Chief marketing officers are always watching the bottom line and gauging their return on investment for every spending decision, and many are forced to achieve these goals within budget constraints. No organization has unlimited marketing resources, so managers are forced to prioritize promotional spending decisions. The optimization models can identify RFM segments that should be reached and RFM segments that are not worthy of pursuing because they are unprofitable or because of budget constraints. The models create a balance between Type I and Type II errors. A Type I error would occur when organizations ignore customers who should have been contacted because they could have returned and repurchased. A Type II error happens when organizations reach customers who are not ready to purchase.[60] The models discussed in this chapter can help identify which customer segments will be most profitable and determine whether to continue promotional spending in an attempt to generate future sales, or whether to curtail spending and allocate those marketing resources to other, more profitable customer targets.

Companies use RFM analysis to determine whether and how to invest in their direct marketing customers. The RFM approach is often used as a promotional decision-making tool in which "promotional spending is allocated on the basis of people's amount of purchases and only to a lesser degree on the basis of their lifetime of duration."[61]

Recency Value

Recency refers to the time of a customer's most recent purchase. Recency is considered especially important because a relatively long period of purchase inactivity can signal to the firm that the customer has ended the relationship. Recency values are assigned to each customer and these values represent the following categories on a scale from 1 to 5:

1. Not recent at all
2. Not recent
3. Somewhat recent
4. Recent
5. Very recent

The specific cutoff points for each category and the number of categories depend on the specific marketing campaign and are decided by the marketing team based on the type of purchase. For example, car dealers may consider a recent purchase as any purchase within the last three years, whereas a grocery store may consider a recent purchase as any purchase during the last week.

Frequency Value

Frequency is defined as the number of a customer's past purchases. Frequency values are assigned to each customer and these values represent the following categories on a scale from 1 to 5:

1. Not frequent at all
2. Not frequent
3. Somewhat frequent
4. Frequent
5. Very frequent

The specific cutoff points for each category and the number of frequency categories are decided by the marketing team based on the type of purchase. An approach to identifying the cutoff points is to simply split the customers equally among the five categories. For example, if the lowest frequency is 50 purchases and the highest frequency is 250 purchases, then the difference of 200 can be divided by

5 resulting in a difference of 40 among the groups. The cutoff point will then be assigned as shown in Table 8-1.

Table 8-1 Example of Cutoff Points for Frequency of Purchases

Frequency Value	Range	Description
1	0–90	Not frequent at all
2	90–130	Not frequent
3	131–170	Somewhat frequent
4	171–210	Frequent
5	211 and above	Very frequent

Monetary Value

Monetary value is based on the average purchase amount per customer transaction. Sometimes, monetary value is calculated as the total amount spent by a customer on all purchases over a specified time period. In this chapter, the average amount of purchase is used and categories are defined as:

1. Very small buyer
2. Small buyer
3. Normal buyer
4. Large buyer
5. Very large buyer

The specific cutoff points for monetary value and the number of monetary value categories can also be decided based on the type of purchases. Just as in the case of frequency, using the quintile values for the average price can be an alternative approach for the cutoff points. For example, if the lowest average purchase per customer is $10 and the highest is $110 purchases, then the difference of 100 can be divided by 5 resulting in a difference of $20 among the groups. The cutoff point will then be assigned as shown in Table 8-2.

Table 8-2 Example of Cutoff Points for the Monetary Value

Monetary Value	Range	Description
1	$0–$30	Very small buyer
2	$31–$50	Small buyer
3	$51–$70	Normal buyer
4	$71–$90	Large buyer
5	$91 and above	Very large buyer

RFM has been available for many years as an analytical technique for marketing campaigns. Although more sophisticated methods have been developed recently, RFM continues to be used because of its simplicity. Direct marketing campaigns, in particular, have become more efficient because of the use of such data mining techniques that allow marketers to better segment and manage their customer databases and to generate more effective and cost-efficient promotional strategies that maximize profits derived from customers' responses. For example, marketing managers may launch a new discount pricing campaign to reach those customers who have low recency values but relatively high frequency and monetary values. Similarly, an organization could launch an up-selling campaign to reach those customers who have high recency and frequency values but low monetary values. Alternatively, the department might launch a cross-selling campaign to reach those customers who have high recency and monetary values but low frequency values.

RFM Analysis with Excel

Excel is an excellent tool to download and transform customer transaction records and prepare them for further RFM analysis. Electronic logs for sales can be summarized with a pivot table, and then recency, frequency, and monetary values can be transformed into a 1 to 5 scale with lookup functions. An explanation of pivot table and lookup functions is offered in Appendix A. The following sections discuss the specific steps needed to apply these Excel tools in the case of RFM analysis.

Using a Pivot Table to Summarize Records

An online coffee retailer (OCR) has recorded thousands of transactions during the last two years. OCR's customers consist of individuals who prefer a variety of coffee brands as well as small businesses, which periodically order coffee for their daily office supplies. Figure 8-3 shows the first few rows (24 transactions) from a total of 6,680 online transactions downloaded in an Excel file. To access the whole set of records, the *ch8_OCR_RFM* file can be downloaded from the companion website at www.informit.com/title/9780133760354. For each transaction, the *Customer ID*, *Transaction Date*, and *Sales* are recorded.

	A	B	C
1	Customer ID	Transaction Date	Sales
2	00000320	6/30/2014	$22.88
3	00000767	6/30/2014	$11.57
4	00000549	6/29/2014	$23.58
5	00001533	6/28/2014	$22.49
6	00000810	6/28/2014	$23.99
7	00000968	6/28/2014	$39.48
8	00001502	6/28/2014	$36.48
9	00002312	6/28/2014	$44.99
10	00000211	6/27/2014	$72.47
11	00000499	6/27/2014	$50.97
12	00001056	6/27/2014	$70.39
13	00001387	6/27/2014	$62.36
14	00001533	6/27/2014	$62.08
15	00002031	6/27/2014	$36.98
16	00000529	6/26/2014	$38.97
17	00000767	6/26/2014	$23.58
18	00001203	6/26/2014	$54.36
19	00001470	6/26/2014	$41.98
20	00000382	6/25/2014	$23.99
21	00000713	6/25/2014	$28.49
22	00000899	6/25/2014	$20.49
23	00001712	6/25/2014	$52.55
24	00001899	6/25/2014	$37.48
25	00002175	6/25/2014	$31.99

Figure 8-3 Latest 24 transactions for OCR

Using a pivot table, the modeler can calculate the most recent transaction date as recency for each customer, the number of transactions during the last two years as frequency for each customer, and the average sales as monetary value for each customer. Briefly, the following steps can be followed in the OCR case:

1. Highlight columns A, B, and C.
2. Click on the Insert tab and then on Pivot Table.
3. Choose New Worksheet to place the pivot table and click OK.
4. Drag Customer ID in the Row Labels section of the Pivot Table field list. Name the column Cust ID.
5. Drag Transaction Date in the Values section of the Pivot Table field list. Using the Value Field Settings, select the Max option to indicate only the most recent transaction date for each customer. Name the column Recency and use the Date format for this column.
6. Drag Sales in the Values section of the Pivot Table field list. Using the Value Field Settings, select the Count option to indicate the number of sales or the frequency of sales for each customer. Name the column Frequency and use the Number format for this column.
7. Drag Sales again in the Values section of the Pivot Table field list. Using the Value Field Settings, select the Average option to indicate the average amount of sales or the monetary value for each customer. Name the column Monetary and use the Currency format for this column.

The partial results of the pivot table and simple statistics are shown in Figures 8-4 and 8-5. Figure 8-4 shows the top of the pivot table and Figure 8-5 shows the bottom section along with a few simple statistics. As shown, a total of 2,349 customers have made purchases during the last two years. The latest purchases (recency) range from 1/2/2013 to 6/30/2014; the frequency ranges from 1 to 44 transactions per customer; and the monetary value ranges from $11.00 to $517.97.

	A	B	C	D
3	Cust ID	Recency	Frequency	Monetary
4	00000001	12/12/2013	4	$36.13
5	00000002	1/13/2013	2	$48.56
6	00000006	6/20/2014	15	$82.40
7	00000007	2/5/2013	1	$22.77
8	00000008	3/7/2014	2	$24.38
9	00000009	2/11/2014	4	$38.14
10	00000011	11/11/2013	6	$41.03
11	00000013	5/27/2014	7	$31.52
12	00000017	1/6/2014	12	$37.15
13	00000018	2/4/2013	1	$20.99
14	00000019	5/5/2014	6	$39.35
15	00000020	1/2/2013	1	$28.96
16	00000021	4/14/2013	3	$37.53
17	00000022	1/2/2013	1	$19.77
18	00000023	1/2/2013	1	$25.70
19	00000024	1/2/2013	1	$25.96
20	00000025	12/21/2013	2	$62.37
21	00000026	1/13/2013	2	$126.57
22	00000027	1/2/2013	1	$26.90
23	00000028	7/27/2013	2	$64.30
24	00000029	1/2/2013	1	$57.37
25	00000030	2/12/2013	2	$45.62
26	00000031	1/2/2013	1	$55.77
27	00000032	1/2/2013	1	$21.77

Figure 8-4 Partial top results of the pivot table

2344	00002349	5/29/2013	2	$25.17
2345	00002350	12/15/2013	5	$44.26
2346	00002351	3/25/2013	1	$25.37
2347	00002352	2/22/2014	2	$25.23
2348	00002353	3/30/2014	3	$34.16
2349	00002354	9/11/2013	6	$55.11
2350	00002355	2/1/2014	2	$29.19
2351	00002356	6/7/2014	7	$40.00
2352	00002357	3/25/2013	1	$36.74
2353	(blank)			
2354	Grand Total	6/30/2014	6680	$47.46
2355				
2356	Count	2349		
2357	Min	1/2/2013	1	$11.00
2358	Max	6/30/2014	44	$517.97
2359	Average		2.84	$44.01

Figure 8-5 Bottom part of the pivot table and summary statistics

Using VLOOKUP to Assign RFM Scores

Considering the output from the pivot table, the modeler decides to use the cutoff points shown in Figure 8-6. The recency cutoff points on the left indicate that all customers whose last purchase ranges from 1/1/2013 to 4/1/2013 will be assigned a score of 1. Customers whose last purchase ranges from 4/1/2013 to 7/1/2013 will be assigned a score of 2, and so on. Similarly, frequency cutoff points in the middle indicate that all customers who have purchased between 0 to 3 times during the last two years will be assigned a score of 1. Customers who have purchased between 4 and 6 times will be assigned a score of 2, and so on. Finally, the monetary cutoff points on the right indicate that all customers whose average purchase ranges from $0 to $25 will be assigned a score of 1. Customers whose average purchase ranges from $25.01 to $50 will be assigned a score of 2, and so on.

Recency Cutoffs			Frequency Cutoffs			Monetary Cutoffs	
1/1/2013	1		0	1		$0	1
4/1/2013	2		3	2		$25	2
7/1/2013	3		6	3		$50	3
11/1/2013	4		9	4		$75	4
2/1/2014	5		12	5		$100	5

Figure 8-6 RFM cutoff points

Figure 8-7 shows the application of the VLOOKUP function to assign an RFM score for each customer. Cell E14, for example, contains the VLOOKUP formula used to assign the recency score for customers in row 14. The VLOOKUP function, VLOOKUP (B14, $K$5:$L$9, 2) will select the recency value for this customer (5/5/2014) in cell B14, compare the value to the recency cutoff ranges (K5:L9), and return the value on the second column of this range (5). The dollar signs before letters and numbers in the selected range lock the range and allow the modeler to extend the formula to all customers. The use of VLOOKUP to assign F-scores and M-scores is demonstrated for customers 13 and 12, respectively.

The RFM score calculated can provide useful insight for marketing analysts to design better marketing campaigns. For example, Figure 8-8 shows the distribution of the customers based on each criterion. It is clear that most of the customers of OCR have not purchased very recently during the two-year period. Twenty percent of them are very recent, while 58% of them are not very recent. Also, the frequency and monetary distributions indicate that customers tend not to return and not to spend a lot of money. It is imperative for the Marketing Department to initiate campaigns with the goal of making customers return, purchase more often, and buy more every time they purchase.

	Recency	Frequency	Monetary	R-score	F-Score	M-Score	
	12/12/2013	4	$36.13	4	2	2	
	1/13/2013	2	$48.56	1	1	2	
	6/20/2014	15	$82.40	5	5	4	
	2/5/2013	1	$22.77	1	1	1	
	3/7/2014	2	$24.38	5	1	1	
	2/11/2014	4	$38.14	5	2	2	
	11/11/2013	6	$41.03	4	3	2	
	5/27/2014	7	$31.52	5	3	2	
	1/6/2014	12	$37.15	4	5	2	<----- =VLOOKUP(D12,$Q$5:$R$9,2)
	2/4/2013	1	$20.99	1	5	1	<----- =VLOOKUP(C13,$N$5:$O$9,2)
	5/5/2014	6	$39.35	5	5	2	<----- =VLOOKUP(B14, $K$5:$L$9,2)
	1/2/2013	1	$28.96	1	1	2	
	4/14/2013	3	$37.53	2	2	2	

Recency Cutoffs

1/1/2013	1
4/1/2013	2
7/1/2013	3
11/1/2013	4
2/1/2014	5

Frequency Cutoffs

0	1
3	2
6	3
9	4
12	5

Monetary Cutoffs

$0	1
$25	2
$50	3
$75	4
$100	5

Figure 8-7 Applying VLOOKUP to generate R-F-M scores

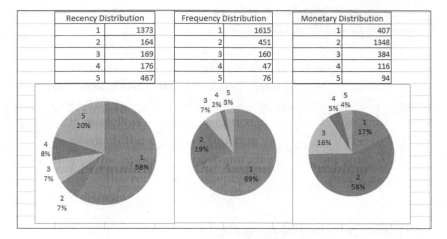

Recency Distribution		Frequency Distribution		Monetary Distribution	
1	1373	1	1615	1	407
2	164	2	451	2	1348
3	169	3	160	3	384
4	176	4	47	4	116
5	467	5	76	5	94

Figure 8-8 Distributions of customers by recency, frequency, and monetary value

Another approach would be to cluster customers based on similar characteristics and apply different marketing campaigns for each cluster. For example, a cluster of customers with low recency scores but with high frequency and monetary scores represents a very valuable group of customers who buy often and spend large amounts of money but may have taken their business to the competitor. The Marketing Department must design a strategy to bring back this group of customers.

Optimizing RFM-Based Marketing Campaigns

The goal of the marketing analyst is to find those values for RFM scores that should be targeted in a given marketing campaign in order to maximize profitability. Simultaneously, the analyst needs to identify those RFM segments that are not worth pursuing either due to unprofitability or due to insufficient campaign budgets. Several linear programming models are offered, each with a special focus on recency, frequency, and monetary value.

LP Models with Single RFM Dimension

This section demonstrates three cases, each with a focus on the respective categories of RFM: recency, frequency, and monetary value. Each case is formulated as a 0-1 LP model, which seeks to identify the category of customers to be reached. In addition, each case is formulated as a continuous LP model, which seeks to identify the portion of customers in each category to be reached. The *ch8_ RFM_with_LP_single_dimension* file is used to illustrate these single dimension models.

LP Model for the Recency Case

Assume that the Marketing Department is able to record the dates of customer transactions and wants to use the recency as the only significant value in its direct marketing campaign. Let the decision variable for this case be a binary variable as follows:

x_i = 1 if customers in recency i are reached through the marketing campaign; 0, otherwise

where:

$i = 1... 5$ index used to identify the group of customers in a given recency category

Also, let:

N_i = number of customers in category i

p_i = probability that a customer of recency i will respond to the campaign

V_i = average amount spent by a customer in recency group i

C = average cost to reach a customer during the marketing campaign

B = available budget for the marketing campaign

These parameters are calculated via the RFM Excel template, as explained in the previous section. The probability that a customer in a given recency group will make a purchase can be calculated using previously recorded data of customer responses from past campaigns. If this data is not available, then the customer response rates (p_i) can be assigned a value of 1 by simply assuming that the customer will respond to the campaign. However, recording and incorporating the response data is highly suggested to increase the accuracy of the proposed models.

Figure 8-9 shows a partial list of records processed via a pivot table and the calculations for the V_i, p_i, and N_i. N_i is calcluated with a COUNTIF function and V_i and p_i are calculated with AVERAGEIF functions. Using the definition of decision variables, a 0-1 mixed integer LP formulation is presented:

$$\text{Maximize: } Z_r = \sum_{i=1}^{5} N_i(p_iV_i - C)x_i \tag{8.1}$$

subject to:

$$\sum_{i=1}^{5} N_iCx_i \leq B \tag{8.2}$$

$$x_i = \{0,1\} \ i = 1 \dots 5 \tag{8.3}$$

Equation (8.1) is the objective function. It maximizes the expected revenue (Z_r) of the marketing campaign. As previously stated, a customer in a state of recency i has a p_i chance of purchasing and a $(1 - p_i)$ chance of not purchasing. When purchasing, the revenue from a customer is calculated as $(V_i - C)$, the difference between the amount of purchases and the average cost to reach the customer. When not purchasing, the expected revenue is simply $(-C)$. As such, the expected value of the revenue generated by a single customer in state i is:

$$p_i(V_i - C) + (1 - p_i)(-C) \tag{8.4}$$

This can be simplified as:

$$p_iV_i - C \tag{8.5}$$

	A	B	C	D	E	F	G	H	I
1	Cust ID	Recency	Frequency	Monetary	R-score	F-Score	M-Score	Response Rate	Sales
2	1	12/12/2013	4	$36.13	4	2	2	0.20	$144.50
3	2	1/13/2013	2	$48.56	1	1	2	0.86	$97.11
4	6	6/20/2014	15	$82.40	5	5	4	0.62	$1,236.05
5	7	2/5/2013	1	$22.77	1	1	1	0.82	$22.77
6	8	3/7/2014	2	$24.38	5	1	1	0.12	$48.76
7	9	2/11/2014	4	$38.14	5	2	2	0.87	$152.57
8	11	11/11/2013	6	$41.03	4	3	2	0.94	$246.18
9	13	5/27/2014	7	$31.52	5	3	2	0.13	$220.65

	K	L	M	N	O	P	Q	R	S	T
1										
2	Recency Cutoffs		Vi	pi	Ni					
3	1/1/2013	1	$49.02	0.50	1373	<----=COUNTIF($E$2:$E$2350,L3)				
4	4/1/2013	2	$171.06	0.50	<----=AVERAGEIF($E$2:$E$2350,L4,$H$2:$H$2350)					
5	7/1/2013	3	$164.62	<----=AVERAGEIF($E$2:$E$2350,L5,$I$2:$I$2350)						
6	11/1/2013	4	$211.57	0.56	176					
7	2/1/2014	5	$335.36	0.51	467					
8										

Figure 8-9 Calculating parameters for the LP recency model

Because there are N_i customers in the recency i, the expected revenue from this group of customers is:

$$N_i(p_i V_i - C) \tag{8.6}$$

As such, (8.6) indicates the sum of revenues for all customers in group i for which a marketing decision to advertise to them ($x_i = 1$) is made. Equation (8.2) ensures that the budget B for this marketing campaign is not exceeded. The left side of the equation represents the actual cost of the campaign, which is calculated as the sum of campaign costs for each group i of customers. Equation (8.3) represents the binary constraints for the decision variables x_i.

Solving the LP Model for the Recency Case

Assume that the marketing campaign for OCR is B = $5,000 and the cost to reach a customer is C = $7.50. Figure 8-10 shows the Solver setup and the final solution for this case. As shown, the company should only select customers of recency 4 and 5 for future promotional efforts. This solution generates net revenue of $95,295. This solution indicates a required budget allocation of only $4,823 compared with the maximum available budget of $5,000. The binary requirement for the decision variables indicates that the company must either reach all customers in a certain recency group or not reach them at all. In the OCR problem, once all customers in the recency groups 4 or 5 are reached, there is no remaining budget to reach *all* customers of another group.

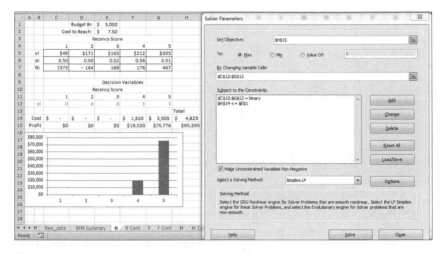

Figure 8-10 Optimal solution for the recency model with 0-1 decision variables

If the modeler changes the requirement for the decision variables to indicate the portion (or percentage) of customers to be reached in each group, then the solution will fully utilize the allocated budget. This solution is shown in Figure 8-11. Notice the change of constraint from C12:G12 = binary to C12:G12 ≤ 1. The solution suggests that the company should reach 100% of the customers with recency 4 and 5 and 14% of the customers with recency 2. This solution will fully utilize the $5,000 budget and provide net revenue of $97,156.

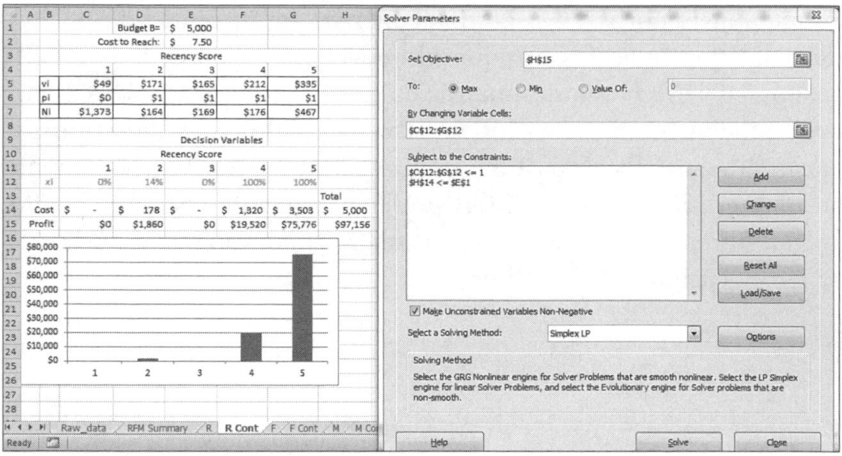

Figure 8-11 Optimal solution for the recency model with continuous decision variables

LP Model for the Frequency Case

This case is relevant to companies where frequency is a significant value in their direct marketing campaign. Let the decision variable be:

x_j = 1 if customers in frequency j are reached through the marketing campaign; 0, otherwise

where:

$j = 1...5$ index used to identify the group of customers in a given frequency category

Also, let:

N_j = number of customers in category j

p_j = probability that a customer of frequency j will respond to the campaign

V_j = average amount spent by a customer in frequency group j

C = average cost to reach a customer during the marketing campaign

B = available budget for the marketing campaign

Table 8-3 shows the calculations of the previous parameters. Just as in the case of recency, N_j is calculated with a COUNTIF function and V_j and p_j are calculated with AVERAGEIF functions. The only difference is that the range has moved from the recency column (E2:E2350) to the frequency column (F2:F2350).

Table 8-3 Parameters for LP Frequency Model

Frequency Cutoffs		V_j	P_j	N_j
0	1	$53.51	0.50	1615
3	2	$169.50	0.51	451
6	3	$341.29	0.53	160
9	4	$495.37	0.50	47
12	5	$1,003.52	0.52	76

Using the definition of decision variables, a 0-1 mixed integer LP formulation is presented:

$$\text{Maximize: } Z_f = \sum_{j=1}^{5} N_j (p_j V_j - C) x_j \quad (8.7)$$

subject to:

$$\sum_{j=1}^{5} N_j C x_j \leq B \quad (8.8)$$

$$x_j = \{0,1\} \; i = 1 \dots 5 \; (8.9) \quad (8.9)$$

Equation (8.7) is the objective function. It maximizes the expected revenue (Z_f) of the marketing campaign. Equation (8.8) ensures that the budget B for this marketing campaign is not exceeded. The left side of the equation represents the actual cost of the campaign, which is calculated as the sum of campaign costs for each group j of customers. Equation (8.9) represents the binary constraints for the decision variables x_j.

Solving the LP Model for the Frequency Case

The marketing campaign has the same budget B = $5,000 and cost to reach a customer C = $7.50. Figure 8-12 shows the Solver Parameters setup and the final solution for this case.

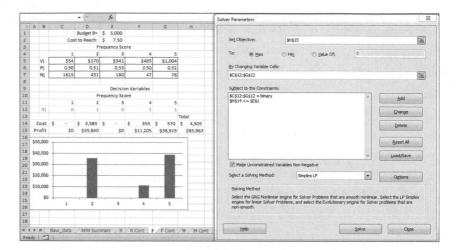

Figure 8-12 Optimal solution for the frequency model with 0-1 decision variables

According to the distribution of customers based on the frequency, the optimal solution suggests that the company should select customers of frequency 2, 4, and 5. As you saw in Table 8.3, the majority of customers are in frequency 1, so the decision to reach this group is not feasible from a cost perspective. This solution indicates revenue of $85,963 and requires a budget allocation of only $4,305 compared with the maximum available budget of $5,000.

If the binary requirement for the decision variables changes to a continuous variable to allow partial reach of a given group, then the solution will fully utilize the allocated budget. This solution is shown in Figure 8-13. The solution suggests that the company should reach 100% percent of the customers with frequency 3, 4, and 5 and 85% of the customers with frequency 2. This solution will fully utilize the $5,000 budget and provide net revenue of $108,442.

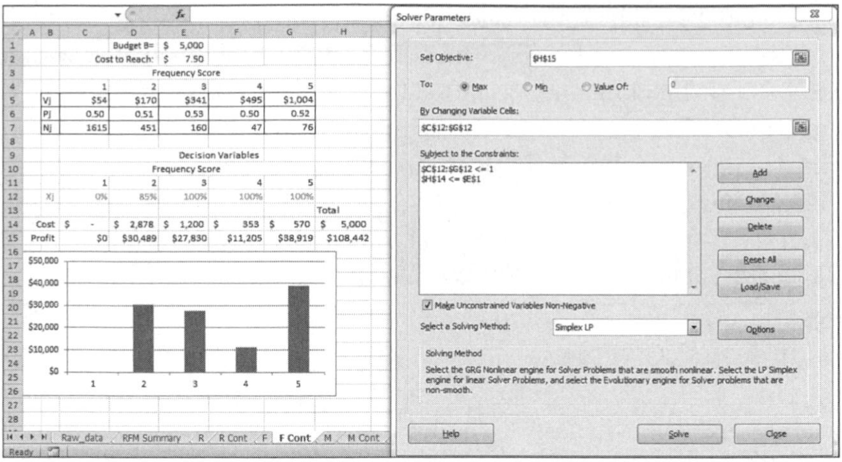

Figure 8-13 Optimal solution for the frequency model with continuous decision variables

LP Model for the Monetary Value Case

This case focuses on the monetary value for each customer. Let the decision variable be:

$x_k = 1$ if customers in monetary group k are reached through the marketing campaign; 0, otherwise

where:

$k = 1...5$ index used to identify the group of customers in a given monetary value category

Also, let:

N_k = number of customers in category k

p_k = probability that a customer of category k will respond to the campaign

V_k = average amount spent by a customer in category j

C = average cost to reach a customer during the marketing campaign

B = available budget for the marketing campaign

Table 8-4 shows the calculations of the previous parameters. Just as in the cases of recency and frequency, N_k is calcluated with a COUNTIF function and V_k and p_k are calculated with AVERAGEIF functions. The only difference is that the range has moved from the frequency column (F2:F2350) to the monetary value column (G2:G2350).

Table 8-4 Parameters for LP Monetary Model

Monetary Cutoffs		V_k	P_k	N_k
$0	1	$32.57	0.52	407
$25	2	$111.92	0.50	1348
$50	3	$203.20	0.50	384
$75	4	$293.81	0.51	116
$100	5	$433.88	0.49	94

Using the definition of decision variables, a 0-1 mixed integer LP formulation is presented:

$$\text{Maximize: } Z_m = \sum_{k=1}^{5} N_k (p_k V_k - C) x_k \tag{8.10}$$

subject to:

$$\sum_{k=1}^{5} N_k C x_k \leq B \tag{8.11}$$

$$x_k = \{0,1\} \; i = 1 \dots 5 \tag{8.12}$$

Equation (8.10) is the objective function. It maximizes the expected revenue (Z_m) of the marketing campaign. Equation (8.11) assures that the budget B for this marketing campaign is not exceeded. The left side of the equation represents the actual cost of the campaign, which is calculated as the sum of campaign costs for each group k of customers. Equation (8.12) represents the binary constraints for the decision variables x_k.

Solving the LP Model for the Monetary Value Case

The marketing campaign has the same budget B = $5,000 and cost to reach a customer C = $7.50. Figure 8-14 shows the Solver Parameters setup and the final solution for this case.

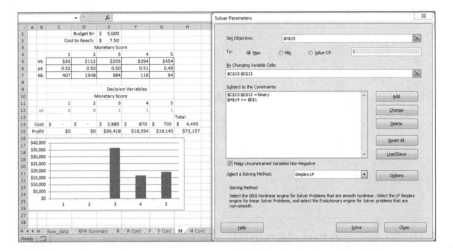

Figure 8-14 Optimal solution for the monetary model with binary decision variables

According to the distribution of customers based on the frequency, the optimal solution suggests that the company should select customers of monetary group 3, 4, and 5. This solution indicates revenue of $72,157 and requires a budget allocation of $4,455. If continuous variables are allowed, then the solution will fully utilize the allocated budget. The solution with continuous decision variables is shown in Figure 8-15 and suggests that the company should reach 100% of the customers with frequency 3, 4, and 5 and 5% of the customers with frequency 2. This solution will fully utilize the $5,000 budget and provide net revenue of $75,698.

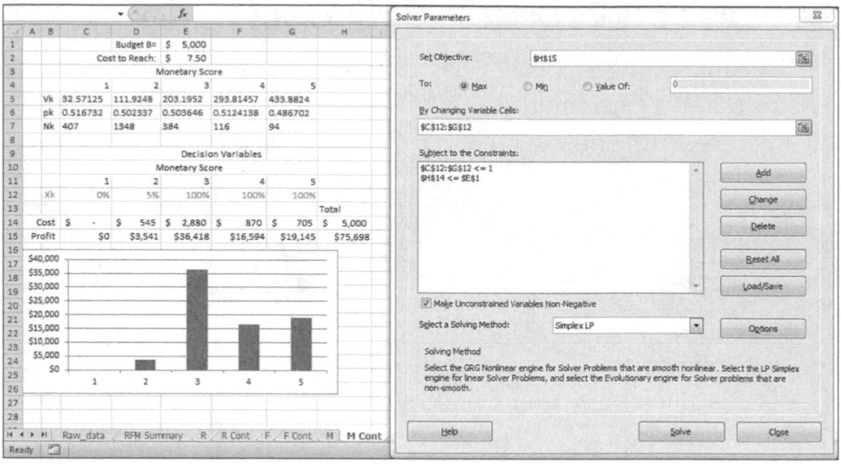

Figure 8-15 Optimal solution for the monetary model with continuous decision variables

Marketing Analytics and Big Data

As mentioned at the start of this chapter, marketing analytics is a component of business analytics. Therefore, many challenges and opportunities of implementing marketing analytics in the era of Big Data are similar to those of business analytics. However, Big Data marketing analytics tend to be mostly generated by customers in the form of structured data from sales transactions and unstructured data from social media networks. Under these circumstances, the decision makers tend to have less control over data-driven decisions. The results of marketing models are driven by the accuracy of data and also by other market forces, especially by the competitors' reactions. For example, if a decision is made to launch an online advertisement campaign but the competition cuts prices by 40%, the actual results of your data-driven decision will not be the same unless "your models account for those sorts of competitive realities."[62] As such, business analytics in general and marketing analytics in particular must use "the right mix of intuition and data-driven analysis" as the ultimate key to success in the era of Big Data.[63]

A successful marketing analytics project requires a supportive analytics culture, support from the top management team, appropriate data, analytics skills, and necessary information technology support. [64] As demonstrated throughout this book, business and marketing analysts must use practical approaches to building models with relatively simple tools, such as Excel.[65]

Wrap Up

This chapter combined the simplicity and intuition of the RFM approach with descriptive, predictive, and prescriptive marketing analytics. The road map to this analysis is summarized in Figure 8-16.

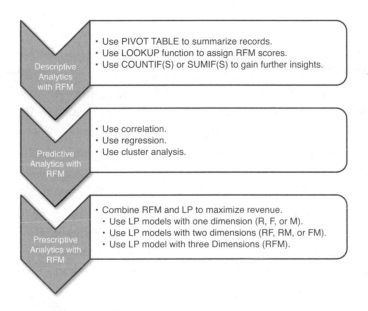

Figure 8-16 Overview of marketing analytics with RFM

RFM analysis has been a popular method in industry to conduct direct marketing campaigns. Marketing analysts prefer the simplicity of the method and its reliance on three important metrics: recency, frequency, and monetary value. RFM can be used as a successful descriptive analytics approach for identifying the value of customers and placing them into more homogenous segments. Each segment is then approached with a unique marketing strategy, which better fits the characteristics of the targeted segment.

Review Questions

1. Is there a difference between business analytics and marketing analytics? Select a real business and provide examples of applications of business analytics and marketing analytics.

2. Just as in business analytics, the domain of marketing analytics includes descriptive, predictive, and prescriptive analysis. Provide examples of each type of analytics using your workplace or a business where you are a regular customer.

3. How can an RFM approach be used for descriptive analytics? What Excel function can you use to generate insight with RFM-based descriptive analytics?

4. How can an RFM approach be used for predictive analytics? What Excel function can you use to generate insight with RFM-based predictive analytics?

5. How can an RFM approach be used for prescriptive analytics? What Excel function can you use to generate insight with RFM-based prescriptive analytics?

6. Explain the concept of customer lifetime value and how it can be incorporated in the LP model. How can an RFM approach be incorporated in the LP model?

7. Explain the difference between Type I and Type II errors when deciding the segment of customers to be reached in a given advertisement campaign. Discuss how the models in this chapter can offer a balance between these two types of errors. Provide examples to illustrate your ideas.

8. How can the RFM score be used to generate more effective campaigns? For example, what type of campaign should be implemented for a group of customers who have low recency values but high frequency and monetary values? What about customers who have high recency and frequency values but low monetary values? What about customers who have high recency and monetary values but low frequency values?

9. What is a pivot table and how can it be used to summarize the records according to recency, frequency, or monetary values? What is the role of the VLOOKUP function in the RFM analysis?

10. Cutoff points for recency, frequency, and monetary values are important in the RFM analysis. What factors should the data analyst consider when determining the cutoff points?

11. The LP models discussed in this chapter use the probability that a customer in a given group will actually return/purchase as a result of the campaign. Assuming that the company does not have the data to calculate the probability of return, can the LP models still be used? If yes, what adjustment would you make in the LP formulation model?

12. The LP models used in this chapter are first solved with a constraint where decision variables must be binary values. Then, the assumption is changed to decision variables as continuous values. Conceptually, what is the difference and what should the data scientist expect in the solution when solving both situations?

13. What are some challenges posed by Big Data for marketing analytics? How can organizations make sure that marketing analytics projects succeed?

Practice Problems

1. You are the marketing analyst for a local restaurant. The restaurant periodically conducts advertising campaigns that include offers for gift cards, free appetizers, and priority seating. The restaurant has established a membership system and has assigned a passport number to its members and is able to track dining history and the amount spent at each visit. Download the file named *ch8_P1restaurant* from the companion website at www.informit.com/title/9780133760354. The file contains 18,095 transactions over the last two years. Each record includes the transaction number, the date of visit, the amount of bill, and the member's passport number.

   a. Use a pivot table to identify the most recent transaction, frequency, and monetary value for each passport member.

   b. Use lookup functions to assign scores for the recency, frequency, and monetary values. The following cutoff points can be used:

Recency Cutoffs		Frequency Cutoffs		Monetary Value Cutoffs	
9/1/2012	1	1	1	$0	1
5/1/2013	2	10	2	$10	2
9/1/2013	3	20	3	$40	3

    c. Generate descriptive statistics using charts, tables, averages, medians, modes, or other summary statistics.

    d. Generate predictive statistics using correlations, regressions, or other forecasting techniques.

    e. Provide insights and marketing recommendations for the company.

2. Continue the analysis of the restaurant from problem 1. You are given $13,000 to conduct the next marketing campaign directed specifically to passport members. Based on the past experiences, it is estimated the average cost to provide incentives to members included in the campaign is $5.75.

    a. Formulate and solve a 0-1 LP model that uses recency data and seeks to maximize the expected revenue under budget limitations. The model should identify the recency values of members who should be included in the campaign.

    b. Adjust the LP model so it identifies the percentage of customers from each recency segment who should be included in the campaign.

    c. Now focus the analysis on the frequency of visits by members. Formulate and solve a 0-1 LP model to identify the frequency segments of customers who should be included in the campaign in order to maximize the expected revenues under the same budget constraints. Also, adjust the model to identify the percentage of customers from each segment.

    d. Finally, focus the analysis on how much members have spent in each visit. Formulate and solve a 0-1 LP model to identify the monetary value segments of customers who should be included in the campaign in order to maximize the expected revenues under the same budget constraints. Also, adjust the model to identify the percentage of customers from each segment.

e. Provide a side-by-side comparison of the results from the above LP models. The results should include overall campaign spending and expected revenue. Provide recommendations to restaurant management regarding the importance of recency, frequency, and monetary values for the members, and segments of customers who should be reached to maximize the future campaign revenues.

3. The Marketing and Sales departments of a shoe store have decided to perform an RFM analysis to determine whether and how to invest in their direct marketing customers. Sales transaction data has been acquired for an entire year. You are the marketing analyst for the shoe store and would like to determine the customer segments to target in an upcoming marketing campaign based on the customers' monetary spending, their recency of purchases, and their frequency. Use the *ch8_P3shoes* file from the companion website to download transactional data for this company.

a. Use a pivot table to identify the most recent transaction, frequency, and monetary value for each customer.

b. Use lookup functions to assign scores for the recency, frequency, and monetary values. Use the 20th, 40th, 60th, and 80th percentiles to establish cutoff points for each dimension of recency, frequency, and monetary values.

c. Generate descriptive statistics using charts, tables, averages, medians, modes, or other summary statistics.

d. Generate predictive statistics using correlations, regressions, or other forecasting techniques.

e. Provide insights and marketing recommendations for the company.

4. Using the information from problem 3, assume that the next campaign for the company has a total budget of $2,000 and the cost to reach a customer of $9.00.

a. Formulate and solve a 0-1 LP model and a continuous LP model that seeks to maximize the expected revenue under budget limitations. The models should identify the recency segments and the percentage of customers in each recency segment who should be included in the campaign.

b. Focus the analysis on the frequency of visits by members. Formulate and solve a 0-1 LP model and a continuous LP model that seeks to maximize the expected revenue under budget limitations. The models should identify the frequency segments and the percentage of customers in each frequency segment who should be included in the campaign.

c. Now, focus the analysis on monetary spending of customers. Formulate and solve a 0-1 LP model and a continuous LP model that seeks to maximize the expected revenue under budget limitations. The models should identify the monetary value segments and the percentage of customers in each segment who should be included in the campaign.

d. Provide a side-by-side comparison of the results from the preceding LP models. The results should include overall campaign spending and expected revenue. Provide recommendations to the management team regarding the importance of recency, frequency, and monetary values for the customers, and regarding segments of customers who should be reached to maximize the future campaign revenues.

5. A company sells a variety of home decoration items online and it wants to utilize benefits of RFM analysis to identify customer buying patterns to better market its products. The raw sales and transaction data for about 2,000 customers during the last three years has been collected and stored in the *ch8_P5decorations* file, which you can download from the companion website.

a. Use a pivot table to identify the most recent transaction, frequency, and monetary value for each customer.

b. Use lookup functions to assign scores for the recency, frequency, and monetary values. Use the 20th, 40th, 60th, and 80th percentiles to establish cutoff points for recency values, 25th, 50th, and 75th percentiles to establish cutoff points for frequency values, and 10th, 20th, 30th, 40th, 50th, 60th, 70th, 80th, and 90th percentiles to establish cutoff points for the monetary value dimension.

    c. Generate descriptive and predictive statistics and provide insights and marketing recommendations for the company.

6. Using the information from the home decoration exercise in problem 5, assume that the marketing budget is set at $20,000 and the cost to reach each customer with advertising and other incentives is expected to be $6.50.

    a. Formulate and solve three 0-1 LP models, each focused on one of the dimensions: recency, frequency, and monetary value. Each model should seek to identify the segment of customers to be reached to maximize the expected revenue under budget limitations. Compare the results (expected spending and revenue) of the three models and provide recommendations about future marketing campaigns.

    b. Formulate and solve three continuous LP models, each focused on one of the dimensions: recency, frequency, and monetary value. Each model should seek to identify the percentage of customers in each segment be reached to maximize the expected revenue under budget limitations. Compare the expected revenue of the three models and provide recommendations about future marketing campaigns.

7. A hotel resort has been experiencing recent difficulties regarding customer visits in its peak season. The management team at the hotel has decided that a marketing campaign needs to determine which previous customers should be targeted in order to get the most return for each marketing dollar spent. Data about past customer stays has been recorded and is saved in *ch8_P7resort* file, which you can download from the companion website. The following cutoff points are also established:

Recency Cutoffs	
2/1/2009	1
2/1/2011	2
2/1/2012	3
2/1/2013	4
2/1/2014	5

Frequency Cutoffs	
0	1
4	2
6	3
8	4
10	5

Monetary Value Cutoffs	
$0	1
$100	2
$500	3
$1,000	4
$1,500	5

    a. Use a pivot table to identify the most recent stay, frequency of stay, and average amount of the hotel bill for each customer.

    b. Use lookup functions to assign scores for the recency, frequency, and monetary values.

    c. Generate descriptive and predictive statistics and provide insights and marketing recommendations for the hotel.

8. Using the same records as in problem 7, assume that the resort only wants to spend a maximum of $2,000 on a marketing campaign. The cost to reach each customer is $4.50 and the response rate for each customer has been given in the same data file.

    a. Formulate and solve three 0-1 LP models, each focused on one of the dimensions: recency of stay, frequency of stay, and money spent in each stay. Each model should seek to identify the segment of resort guests to be reached to maximize the expected revenue under budget limitations. Compare the results (expected spending and revenue) of the three models and provide recommendations about future marketing campaigns.

b. Formulate and solve three continuous LP models, each focused on one of the same dimensions: recency, frequency, and monetary value. Each model should seek to identify the percentage of guests in each segment be reached to maximize the expected revenue under budget limitations. Compare the expected revenue of the three models and provide recommendations about future marketing campaigns.

9

Marketing Analytics with Multiple Goals

Chapter Objectives

- Discuss the importance of seeking multiple goals in a marketing campaign
- Demonstrate the process of formulating linear models with a combination of two and three dimensions of the RFM approach
- Demonstrate the process of formulating goal programming models with assigned priorities to each dimension of the RFM approach
- Demonstrate the use of Solver for solving goal programming models as a series of several linear programming models
- Discuss the implications of combining mathematical programming models with the RFM approach

Prescriptive Analytics in Action: First Tennessee Bank Improves Marketing Campaigns[1]

First Tennessee Bank is a full-service provider of financial products and services for businesses and consumers. Just like almost all other banks, First Tennessee offers a diverse portfolio of services and

---

[1] IBM Corporation, "First Tennessee Bank: Analytics Drives Higher ROI from Marketing Programs," 2011. [Online]. Available: http://www.ibm.com/smarterplanet/us/en/leadership/firsttenbank/assets/pdf/IBM-firstTennBank.pdf. [Accessed June 11, 2014].

generates a large amount of data from both online and offline transac-tions. The availability of large amounts of data has given First Tennes-see an opportunity to better tailor its marketing strategies. The goal is to "shift from the 'marketing-as-an expense' mindset to the idea that marketing is a true profit driver."[66]

Using predictive analytics, the marketing team was able to utilize lots of customer data points and predict the likelihood to respond to an offer and find the areas for cross-selling opportunities. However, that was only the beginning. "What sets the First Tennessee approach apart is how it applies a rigorous, systematic approach to prioritizing which opportunities make it to the campaign stage."[66]

Advanced marketing models at First Tennessee focus on product revenue and cost information generated from its data warehouse sys-tems. The marketing analysts are able to generate quantitative mea-sures of the expected profits from a specific segment of its customers. Optimal marketing campaigns have led to a 20% reduction in mailing cost and 17% reduction in printing cost while increasing the rate of response by 3.1%. "Overall, the bank has tallied a 600 percent return on its investment in predictive analytics through more efficiently deployed resources."[66]

Introduction

This chapter discusses RFM-based optimization models with multiple objectives. Specifically, the chapter expands on the single-goal RFM-based models discussed in Chapter 8, "Marketing Analyt-ics with Linear Programming," and introduces the following new set of models. The same example from the online coffee retailer (OCR) is used here as well to demonstrate the proposed models:

- Three two-dimensional RFM LP models, which combine any two dimensions, such as RF, RM, and FM

- A three-dimensional RFM LP model, which combines all three dimensions in one LP model

- An RFM-based goal programming (RFM GP) model, which incorporates all three dimensions of the RFM analysis, but assigns different weights to each of them

LP Models with Two RFM Dimensions

This section demonstrates three cases, each with a focus on a combination of two RFM: recency and frequency, recency and monetary value, and frequency and monetary value. Each case is formulated as a 0-1 LP model and as a continuous LP model. The *ch9_RFM_LP_ two_dimensions* file, which you can download from the companion website at www.informit.com/title/9780133760354, is used to illustrate these two-dimensional LP RFM models.

LP Model for the Recency and Frequency Case

This case is relevant to companies where recency and frequency are the only significant values in their direct marketing campaign. In this situation, customers are first organized into *RxF* groups, with each iso-value group G_{ij} containing customers who belong to recency value i (1, 2..., R) and frequency value j (1, 2..., F). Companies are interested in determining which customer groups should be targeted and which groups should not be reached. The objective, again, is to maximize the expected revenues from potential customer purchases while not exceeding the budget constraints.

Let the decision variable for this case be a 0-1 unknown variable as follows:

x_{ij} = 1 if customers in recency i and frequency j are reached in the marketing campaign; 0 otherwise

The 0-1 LP formulation is presented for the recency and frequency case:

$$\text{Maximize: } Z_{rf} = \sum_{i=1}^{R} \sum_{j=1}^{F} N_{ij}(p_{ij}V_{ij} - C)x_{ij} \qquad (9.1)$$

subject to:

$$\sum_{i=1}^{R} \sum_{j=1}^{F} N_{ij}Cx_{ij} \leq B \qquad (9.2)$$

$$x_{ij} = \{0,1\}\ i = 1 \dots R\ j=1\dots F \qquad (9.3)$$

where:

- $i = 1... R$ index is used to identify the group of customers in a given recency category
- $j = 1... R$ index is used to identify the group of customers in a given frequency category
- N_{ij} = number of customers in recency category i and frequency category j
- p_{ij} = probability that a customer of recency i and frequency j will respond to the campaign
- V_{ij} = average amount spent by a customer in recency category i and frequency category j
- C = average cost to reach a customer during the marketing campaign
- B = available budget for the marketing campaign

Equation (9.1) is the objective function, which maximizes the expected revenue (Z_{rf}) of the marketing campaign. As previously stated, a customer in a state of recency i and frequency j has a p_{ij} chance of purchasing and a $(1 - p_{ij})$ chance of not purchasing. When purchasing, the expected revenue from a customer is calculated as $(V_{ij} - C)$. When not purchasing, the expected revenue is simply $(-C)$. As such, the expected value of the revenue from a single customer in state ij is:

$$p_{ij}(V_{ij} - C) + (1 - p_{ij})(-C) \qquad (9.4)$$

This can be simplified as:

$$p_{ij}V_{ij} - C \qquad (9.5)$$

Because there are N_{ij} customers with recency i and frequency j, the expected revenue from this group of customers is:

$$N_{ij}(p_{ij}V_{ij} - C) \qquad (9.6)$$

As such, (9.6) indicates the sum of expected revenue customers in group ij for which a marketing decision to reach them ($x_{ij} = 1$) is made. Equation (9.2) ensures that the budget B for this marketing campaign is not exceeded. The left side of the equation represents the actual cost of the campaign, which is calculated as the sum of campaign costs for each group ij of customers. Equation (9.3) represents the binary constraints for the decision variables x_{ij}.

Solving the LP Model for the Recency and Frequency Case

The parameters of the LP model are calculated using the same RFM Excel template used for the single-dimension LP models. The only difference is the calculation of the probability that a customer in a given recency group and given frequency group will make a purchase. This probability is calculated using AVERAGEIFS functions, which allow for two conditions to be satisfied. The same function can be used to calculate the average amount spent by a customer in recency category i and frequency category j (V_{ij}). The number of customers in recency category i and frequency category j (N_{ij}) is calculated using the COUNTIFS function. These calculations can be found in Figure 9-1. An IFERROR formula encapsulates the AVERAGEIFS function applied for these parameters. Certain combinations of recency and frequency may have no customers, and as such, calculating the average revenue or probability of return for these groups generates a division by zero error. A value of 0 is assigned to those cells using the construct, as illustrated in cells M3, R5, and X7:

M3 = IFERROR(AVERAGEIFS($I$2:$I$2350,$E$2: $E$2350,$L3,$F$2:$F$2350,M$2),0)

R5 = IFERROR(AVERAGEIFS($H$2:$H$2350,$E$2: $E$2350,$L5,$F$2:$F$2350,R$2),0)

X7 = IFERROR(COUNTIFS($E$2:$E$2350,$L7,$F$2: $F$2350,W$2),0)

	A	B	C	D	E	F	G	H	I
1	Cust ID	Recency	Frequency	Monetary	R-score	F-Score	M-Score	Response Rate	Purchases
2	1	12/12/2013	4	$36.13	4	1	2	0.20	$144.50
3	2	1/13/2013	2	$48.56	4	2	2	0.86	$97.11
4	6	6/20/2014	15	$82.40	5	2	4	0.62	$1,236.05
5	7	2/5/2013	1	$22.77	4	1	1	0.82	$22.77
6	8	3/7/2014	2	$24.38	5	1	1	0.12	$48.76
7	9	2/11/2014	4	$38.14	5	1	2	0.87	$152.57
8	11	11/11/2013	6	$41.03	4	1	2	0.94	$246.18

Recency Cutoffs / Frequency: Vij

	K	L	M (1)	N (2)	O (3)	P (4)	Q (5)
11	Recency Cutoffs						
13	1/1/2013	1	$49.02	$0.00	$6,794.70	$0.00	$0.00
14	4/1/2013	2	$130.42	$0.00	$0.00	$0.00	$0.00
15	7/1/2013	3	$157.56	$555.19	$0.00	$0.00	$0.00
16	11/1/2013	4	$184.54	$731.39	$2,207.58	$0.00	$0.00
17	2/1/2014	5	$220.45	$689.65	$934.77	$1,325.23	$1,937.04

N13: <--=IFERROR(AVERAGEIFS($I$2:$I$2350,$E$2:$E$2350,$L13,$F$2:$F$2350,M$2),0)

Frequency: pij

	R (1)	S (2)	T (3)	U (4)	V (5)
13	0.50	-	0.68	-	-
14	0.51	-	-	-	-
15		-	-	-	-
16	0.60	0.85	1	-	-
17	0.48	0.47	0.46	0.81	-

S15: <--=IFERROR(AVERAGEIFS($H$2:$H$2350,$E$2:$E$2350,$L15,$F$2:$F$2350,R$2),0)

Frequency: Nij

	W (1)	X (2)	Y (3)	Z (4)	AA (5)
13	1,373	-	-	1	-
14	163	-	-	-	-
15		-	-	-	-
16	170	5	1	-	-
17	376				

X17: <--=IFERROR(COUNTIFS($E$2:$E$2350,$L17,$F$2:$F$2350,W$2),0)

	A	B					
20	27	1/2/2013	$26.90	$26.90	1	1	2
21	28	7/27/2013	$64.30	$128.60	4	1	3
22	29	1/2/2013	$57.37	$57.37	1	1	3
23	30	2/12/2013	$45.62	$91.23	4	1	3
24	31	1/2/2013	$55.77	$55.77	1	1	3

Figure 9-1 Calculating parameters for the LP recency and frequency model

Figure 9-2 shows the overall template for the RF model. Decision variables are located in the cells L5:P9 and are initially set to 1. The cost of the advertising campaign is calculated as a product of the number of customers in a certain group, the decision variable value, and the cost per customer. For example, the cost in cell L13 is calculated with the following formula:

$$L13 = \$E\$2 \times C21 \times L5$$

where:

$E$2 holds cost per customer (=$7.5)

C21 holds number of customers in recency 1 and frequency 1 (=1373)

L5 holds the value 1 indicating that initially a decision is made to reach customers in recency 1 and frequency 1

This formula is expended to all recency and frequency groups and these values are held in the L13:P17 range. The total initial cost is shown to be $17,587. Obviously, this cost is over the budget ($5,000) as such, and as expected, the initial solution is not feasible.

The expected revenue is calculated using the following formula:

Expected revenue = (number of customers to reach) × (average revenue per customer) × (probability that customer will respond) × (decision whether customer is reached (1) or not reached (0)) – (total cost for this group of customers)

	Vij	Budget B=	$ 5,000			
		Cost to Reach:	$ 7.50			
				Recency Score		
		1	2	3	4	5
Frequency	1	$49	$130	$158	$185	$220
	2	$0	$0	$555	$731	$690
	3	$0	$6,795	$0	$2,208	$935
	4	$0	$0	$0	$0	$1,325
	5	$0	$0	$0	$0	$1,937

	Pij			Recency Score		
		1	2	3	4	5
Frequency	1	0.50	0.50	0.51	0.56	0.51
	2	0.00	0.00	0.80	0.60	0.48
	3	0.00	0.68	0.00	0.85	0.47
	4	0.00	0.00	0.00	0.00	0.46
	5	0.00	0.00	0.00	0.00	0.81

	Nij			Recency Score		
		1	2	3	4	5
Frequency	1	1373	163	166	170	376
	2	0	0	3	5	69
	3	0	1	0	1	14
	4	0	0	0	0	4
	5	0	0	0	0	4

Decision Variables	Xij			Recency Score		
		1	2	3	4	5
Frequency	1	1	1	1	1	1
	2	1	1	1	1	1
	3	1	1	1	1	1
	4	1	1	1	1	1
	5	1	1	1	1	1

Cost of Campaign				Recency Score		
		1	2	3	4	5
Frequency	1	$10,298	$1,223	$1,245	$1,275	$2,820
	2	$0	$0	$23	$38	$518
	3	$0	$8	$0	$8	$105
	4	$0	$0	$0	$0	$30
	5	$0	$0	$0	$0	$0
					Total Budget	$17,587.50

Total Revenue				Recency Score		
		1	2	3	4	5
Frequency	1	$23,071	$9,456	$12,221	$16,188	$39,348
	2	$0	$0	$1,310	$2,171	$22,544
	3	$0	$4,613	$0	$1,869	$6,083
	4	$0	$0	$0	$0	$2,382
	5	$0	$0	$0	$0	$0
						$141,258

Figure 9-2 Initial setup with decision variables, cost, and expected revenue

For example, expected revenue in cell L21 is calculated with the following formula:

$$L21 = =C5 \times C13 \times C21 \times L5 - L13$$

where:

C5 holds revenue per customer in recency 1 and frequency 1 (=$49)

C13 holds the probability that the customer will respond (=0.50)

C21 holds number of customers in recency 1 and frequency 1 (=1373)

L5 holds the value 1 indicating that initially a decision is made to reach customers in recency 1 and frequency 1

L13 holds the cost for the customer group, as shown in the previous paragraph

This formula is expended to all recency and frequency groups and these values are held in the L21:P25 range. The total initial expected revenue is shown to be $141,258.

Figure 9-3 shows Solver parameters for the model. The goal is to maximize the expected revenue (P26) by changing binary variables L5:P9 under the budget constraints P18 <= E1.

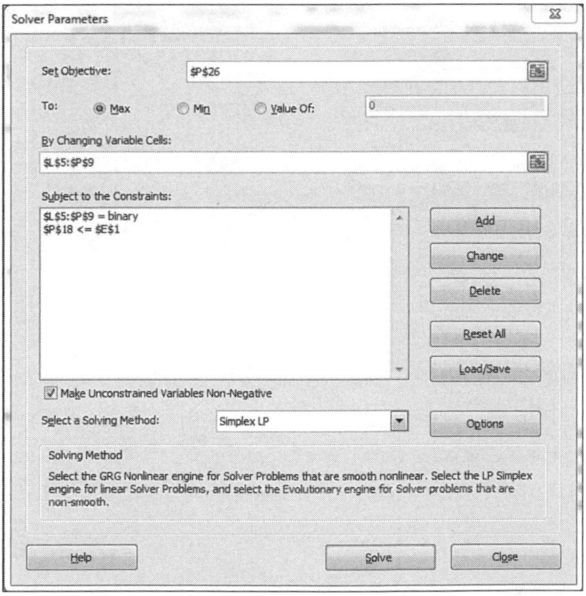

Figure 9-3 Solver Parameters setup for the recency-frequency model

The results demonstrated in Figure 9-4 indicate that the company should only target the following groups of customers: G_{23}, G_{32}, G_{41}, G_{42}, G_{43}, G_{51}, G_{52}, G_{53}, and G_{54} (where G_{ij} represents customer groups of recency i and frequency j). The graph inside the Excel spreadsheet indicates visually the expected revenue of various customer groups as a function of recency and frequency. This solution generates an expected revenue of \$96,508 and requires a total budget \$4,822.

The preceding solution assumes that a certain G_{ij} is either included or not included in the marketing campaign. Often, the analyst wants to determine what percentage of customers in a certain G_{ij} group should be included in the campaign. In this case, the binary decision variables constraints in the Solver Parameters dialog box (\$L\$5:\$P\$9 = binary) are changed to continuous decision variables constraints (\$L\$5:\$P\$9 ≤ 1). This change, in addition to non-negativity constraints, limits the decision variables to a number between zero and one, indicating in essence the percentage of customers in the group to be reached. When implementing this change to the model, the solution shown in Figure 9-5 indicates that the available budget \$5,000 is fully used for maximum expected revenue of \$98,251. This solution suggests that the company should reach 14% of customers in group G_{31} and 100% of customers in groups G_{23}, G_{32}, G_{41}, G_{42}, G_{43}, G_{51}, G_{52}, G_{53}, and G_{54}.

Budget B= $ 5,000
Cost to Reach: $ 7.50

Vij — Recency Score

Vij	1	2	3	4	5
1	$49	$130	$158	$185	$220
2	$0	$0	$555	$781	$690
3	$0	$6,795	$0	$2,208	$935
4	$0	$0	$0	$0	$1,325
5	$0	$0	$0	$0	$1,937

Frequency (rows)

Decision Variables Xij — Recency Score

Xij	1	2	3	4	5
1	0	0	0	1	1
2	0	0	0	1	1
3	0	0	1	1	1
4	0	0	0	0	1
5	0	0	0	0	0

Frequency (rows)

Pij — Recency Score

Pij	1	2	3	4	5
1	0.50	0.50	0.51	0.56	0.51
2	0.00	0.00	0.80	0.60	0.48
3	0.00	0.68	0.00	0.85	0.47
4	0.00	0.00	0.00	0.00	0.46
5	0.00	0.00	0.00	0.00	0.81

Frequency (rows)

Cost of Campaign — Recency Score

Cost of Campaign	1	2	3	4	5
1	$0	$0	$0	$1,275	$2,820
2	$0	$0	$23	$38	$518
3	$0	$8	$0	$8	$105
4	$0	$0	$0	$0	$30
5	$0	$0	$0	$0	$0

Frequency (rows)

Total Budget $4,822.50

Nij — Recency Score

Nij	1	2	3	4	5
1	1373	163	166	170	376
2	0	0	3	5	69
3	0	1	0	1	14
4	0	0	0	0	4
5	0	0	0	0	4

Frequency (rows)

Total Revenue — Recency Score

Total Revenue	1	2	3	4	5
1	$0	$0	$1,310	$16,188	$39,348
2	$0	$0	$0	$2,171	$22,544
3	$0	$4,613	$0	$1,869	$6,083
4	$0	$0	$0	$0	$2,382
5	$0	$0	$0	$0	$0

$96,508

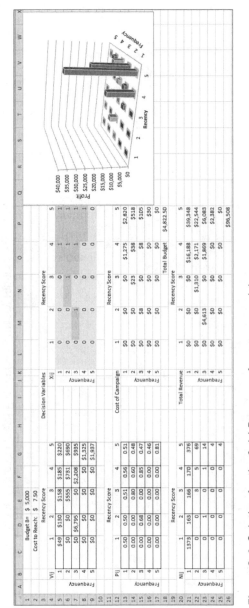

Figure 9-4 Solution for the 0-1 LP recency-frequency model

			Budget B=	$	5,000	
			Cost to Reach:	$	7.5	

Decision Variables — Xij

Recency Score (left)

Frequency	1	2	3	4	5
1	0.50	0.50	0.51	0.56	0.51
2	0.00	0.00	0.80	0.60	0.48
3	0.00	0.68	0.00	0.85	0.47
4	0.00	0.00	0.00	0.00	0.46
5	0.00	0.00	0.00	0.00	0.81

Recency Score (right)

Frequency	1	2	3	4	5
1	0%	0%	0%	100%	100%
2	0%	0%	14%	100%	100%
3	0%	100%	100%	100%	100%
4	100%	0%	0%	0%	100%
5	0%	0%	0%	0%	0%

Cost of Campaign

Recency Score (left)

Frequency	1	2	3	4	5
1	$49	$130	$158	$185	$220
2	$0	$0	$555	$731	$690
3	$0	$6,795	$0	$2,208	$935
4	$0	$0	$0	$0	$1,325
5	$0	$0	$0	$0	$1,937

Recency Score (right)

Frequency	1	2	3	4	5
1	$0	$0	$178	$1,275	$2,820
2	$0	$0	$23	$38	$518
3	$8	$8	$0	$8	$105
4	$0	$0	$0	$0	$30
5	$0	$0	$0	$0	$0

Total Budget	$5,000.00

Total Revenue

Recency Score (left)

Frequency	1	2	3	4	5
1	1373	163	166	170	376
2	0	0	3	5	69
3	0	1	0	1	14
4	0	0	0	0	4
5	0	0	0	0	4

Recency Score (right)

Frequency	1	2	3	4	5
1	$0	$0	$1,742	$16,188	$39,348
2	$0	$0	$1,310	$2,171	$22,544
3	$0	$4,613	$0	$1,869	$6,083
4	$0	$0	$0	$0	$2,382
5	$0	$0	$0	$0	$0
					$98,251

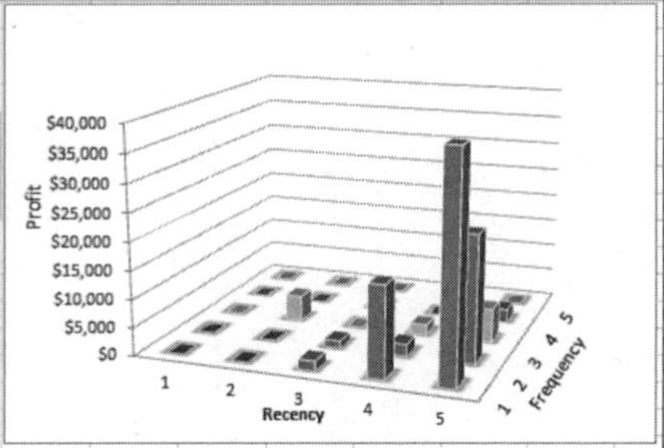

Chart — Profit (axes: Profit $0–$40,000; Recency 1–5; Frequency 1–5)

Figure 9-5 Solution for the continuous LP recency-frequency model

LP Model for the Recency and Monetary Value Case

The approach to the other two-dimensional LP models is very similar. This section uses recency and monetary values as inputs in the marketing campaign. Let the decision variable for this case be a 0-1 unknown variable as follows:

x_{ik} = 1 if customers in recency i and monetary value k are reached in the marketing campaign; 0 otherwise

The 0-1 LP formulation is presented for the recency and monetary value case:

$$\text{Maximize: } Z_{rm} = \sum_{i=1}^{R} \sum_{k=1}^{M} N_{ik}(p_{ik}V_{ik} - C)x_{ik} \qquad (9.6)$$

subject to:

$$\sum_{i=1}^{R} \sum_{k=1}^{M} N_{ik}Cx_{ik} \leq B \qquad (9.7)$$

$$x_{ik} = \{0,1\} \qquad i = 1 \dots R \qquad k = 1 \dots M \qquad (9.8)$$

where:

$i = 1 \dots R$ index is used to identify the group of customers in a given recency category

$k = 1 \dots M$ index is used to identify the group of customers in a given monetary value category

N_{ik} = number of customers in recency category i and monetary value category k

p_{ik} = probability that a customer of recency i and monetary value k will respond to the campaign

V_{ik} = average amount spent by a customer in recency category i and monetary value category k

C = average cost to reach a customer during the marketing campaign

B = available budget for the marketing campaign

Equation (9.6) is the objective function, which maximizes the expected revenue (Z_{rm}) of the marketing campaign. Equation (9.7) ensures that the budget B for this marketing campaign is not exceeded. The left side of the equation represents the actual cost of the campaign, which is calculated as the sum of campaign costs for each group ik of customers. Equation (9.8) represents the binary constraints for the decision variables x_{ik}. For the continuous requirement, this constraint is transformed into:

$$x_{ik} \leq 1 \qquad i = 1 \dots R \qquad k = 1 \dots M$$

Solving the LP Model for the Recency and Monetary Value Case

The Excel template and Solver approach to this case are similar to the recency-frequency case. The same formulas are used to calculate the average revenue, probability of return, and number of customers in each group ik. These calculations can be found in Figure 9-6.

Figure 9-7 shows the final solution of this problem. The results indicate that the company should only target the following groups of customers: G_{23}, G_{24}, G_{33}, G_{34}, G_{42}, G_{43}, G_{44}, G_{52}, G_{53}, and G_{54}. The expected revenue for this solution is \$94,173 and the total budget needed is \$4,927. Just as in the previous models, changing the constraints for the decision variables to be continuous causes the full use of the available budget of \$5,000 and achieves maximum expected revenue of \$94,736. These final values of the decision variables are shown in Figure 9-8.

Figure 9-6 Parameters for LP recency and monetary model

Monetary Cutoffs		Recency: Vij					Recency: pij					Recency: Nij				
		1	2	3	4	5	1	2	3	4	5	1	2	3	4	5
$0	1	$23.37	$56.64	$60.33	$78.54	$88.22	0.52	0.39	0.56	0.49	0.54	330	19	18	15	25
$25	2	$40.22	$100.23	$127.56	$156.69	$271.15	0.49	0.53	0.51	0.57	0.50	737	100	102	111	298
$50	3	$74.54	$168.65	$230.62	$225.84	$443.89	0.50	0.46	0.53	0.45	0.53	196	25	31	30	102
$75	4	$90.85	$249.33	$338.58	$614.35	$561.70	0.49	0.48	0.47	0.74	0.46	56	11	9	14	26
$100	5	$189.83	$1,110.54	$391.87	$548.13	$857.74	0.46	0.53	0.52	0.61	0.48	54	9	9	6	16

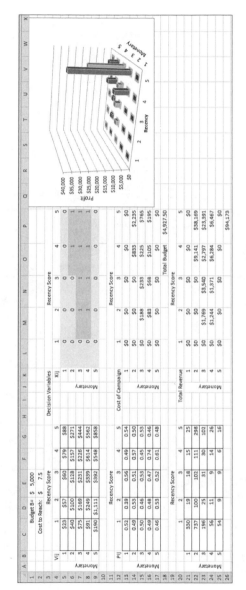

Figure 9-7 Solution for the binary LP recency-monetary model

Budget B= $5,000
Cost to Reach: $7.5

Vij

Monetary	Recency Score 1	2	3	4	5
1	$23	$57	$60	$79	$88
2	$40	$100	$128	$157	$271
3	$75	$169	$231	$226	$444
4	$91	$249	$339	$614	$562
5	$190	$1,111	$392	$548	$858

Pij

Monetary	Recency Score 1	2	3	4	5
1	0.52	0.39	0.56	0.49	0.54
2	0.49	0.53	0.51	0.57	0.50
3	0.50	0.46	0.53	0.45	0.53
4	0.49	0.48	0.47	0.74	0.46
5	0.46	0.53	0.52	0.61	0.48

Monetary	Recency Score 1	2	3	4	5
1	330	19	18	15	25
2	737	100	102	111	298
3	196	25	31	30	102
4	56	11	9	14	26
5	54	9	9	6	16

Decision Variables Xij

Monetary	Recency Score 1	2	3	4	5
1	0%	0%	0%	0%	0%
2	0%	0%	9%	100%	100%
3	0%	100%	100%	100%	100%
4	0%	100%	100%	100%	0%
5	0%	0%	0%	0%	0%

Cost of Campaign

Monetary	Recency Score 1	2	3	4	5
1	$0	$0	$0	$0	$0
2	$0	$0	$73	$833	$2,235
3	$0	$188	$233	$225	$765
4	$0	$83	$68	$105	$195
5	$0	$0	$0	$0	$0

Total Budget $5,000.00

Sum Total Revenue

Monetary	Recency Score 1	2	3	4	5
1	$0	$0	$562	$9,141	$38,169
2	$0	$1,769	$3,540	$2,797	$23,391
3	$0	$1,244	$1,371	$6,284	$6,467
4	$0	$0	$0	$0	$0
5	$0				

$94,736

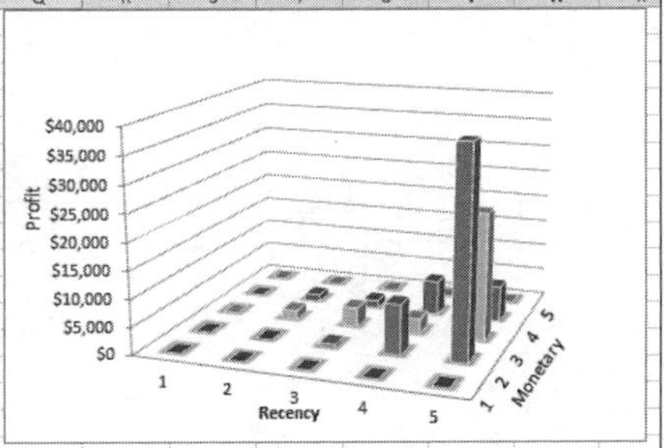

Figure 9-8 Solution for the continuous LP monetary-recency model

LP Model for the Frequency and Monetary Case

Let the decision variable for this case be a 0-1 unknown variable as follows:

x_{jk} = 1 if customers in frequency j and monetary k are reached in the marketing campaign; 0 otherwise

The 0-1 LP formulation is presented for the frequency-monetary case:

$$\text{Maximize: } Z_{fm} = \sum_{j=1}^{F} \sum_{k=1}^{m} N_{jk}(p_{jk}V_{jk} - C)x_{jk} \tag{9.9}$$

subject to:

$$\sum_{j=1}^{F} \sum_{k=1}^{M} N_{jk}Cx_{jk} \leq B \tag{9.10}$$

$$x_{jk} = \{0,1\} \text{ or } x_{jk} \leq 1 \text{ for } j = 1 \dots F \ k=1\dots M \tag{9.11}$$

where:

$j = 1\dots F$ index is used to identify the group of customers in a given frequency category

$k = 1\dots M$ index is used to identify the group of customers in a given monetary category

N_{jk} = number of customers in frequency category j and monetary category k

p_{jk} = probability that a customer of frequency j and monetary k will respond to the campaign

V_{jk} = average amount spent by a customer in frequency category j and monetary category k

C = average cost to reach a customer during the marketing campaign

B = available budget for the marketing campaign

Solving the LP Model for the Frequency–Monetary Case

This case is relevant to companies where frequency and monetary are the only significant values in their direct marketing campaign. Because the approach, formulas, template, and solution are similar to the previous two-dimensional cases, this section simply shows the results of the formulation of the frequency-monetary LP model, parameter calculations (Figure 9-9), binary solution (Figure 9-10), and continuous solution (Figure 9-11).

The solution generates an expected revenue of $76,508 and requires a total budget of $4,230 for the model with 0-1 decision variables and an expected revenue of $80,091 and requires a total budget of $5,000 for the model with continuous decision variables.

Figure 9-9 Parameters for the frequency-monetary model

Frequency Cutoffs		Monetary: Vij					Monetary: pij					Monetary: Nij				
		1	2	3	4	5	1	2	3	4	5	1	2	3	4	5
0	1	$31.78	$84.44	$152.76	$221.26	$320.45	0.52	0.50	0.50	0.51	0.49	406	1,285	359	108	90
10	2	$355.81	$494.77	$817.10	$1,139.87	$1,716.62	0.57	0.49	0.54	0.53	0.46	1	45	21	7	3
20	3	$0.00	$842.80	$1,272.02	$2,207.58	$6,794.70	-	0.50	0.37	0.85	0.68	-	11	3	1	1
30	4	$0.00	$1,325.23	$0.00	$0.00	$0.00	-	0.46	-	-	-	-	4	-	-	-
40	5	$0.00	$1,846.19	$2,209.58	$0.00	$0.00	-	0.86	0.67	-	-	-	3	1	-	-

Figure 9-10 Solution for the binary LP frequency-monetary model

Budget B= $ 5,000
Cost to Reach: $ 7.5

Vij — Frequency Score

Monetary	1	2	3	4	5
1	$32	$356	$0	$0	$0
2	$84	$495	$843	$1,325	$1,846
3	$153	$817	$1,272	$0	$2,210
4	$221	$1,140	$2,208	$0	$0
5	$320	$1,717	$6,795	$0	$0

pij — Frequency Score

Monetary	1	2	3	4	5
1	0.52	0.57	0.00	0.00	0.00
2	0.50	0.49	0.50	0.46	0.86
3	0.50	0.54	0.37	0.00	0.67
4	0.51	0.53	0.85	0.00	0.00
5	0.49	0.46	0.68	0.00	0.00

Nij — Frequency Score

Monetary	1	2	3	4	5
1	406	1	0	0	0
2	1,285	45	11	4	3
3	359	21	3	0	1
4	108	7	1	0	0
5	90	3	1	0	0

Decision Variables — Xij — Frequency Score

Monetary	1	2	3	4	5
1	1	0	0	0	0
2	0	1	1	1	0
3	1	1	1	1	1
4	1	1	1	0	0
5	0	0	0	0	0

Cost of Campaign — Frequency Score

Monetary	1	2	3	4	5
1	$0	$8	$0	$0	$0
2	$0	$338	$83	$50	$23
3	$2,693	$158	$23	$8	$8
4	$810	$53	$8	$0	$0
5	$0	$0	$0	$0	$0

Total Budget: $4,290.00

Total Revenue — Frequency Score

Monetary	1	2	3	4	5
1	$0	$195	$0	$0	$0
2	$0	$10,518	$4,561	$2,382	$4,722
3	$24,857	$9,059	$1,389	$0	$1,473
4	$11,339	$4,142	$1,869	$0	$0
5	$0	$0	$0	$0	$0

$76,508

Budget B= $5,000
Cost to Reach: $ 7.5

Frequency Score — Vij (Monetary)

Vij	1	2	3	4	5
1	$32	$356		$0	$0
2	$84	$495	$843	$1,325	$1,846
3	$153	$817	$1,272	$0	$2,210
4	$221	$1,140	$2,208	$0	$0
5	$320	$1,717	$6,795	$0	$0

Frequency Score — Pij (Monetary)

Pij	1	2	3	4	5
1	0.52	0.57	0.00	0.00	0.00
2	0.50	0.49	0.50	0.46	0.86
3	0.50	0.54	0.37	0.00	0.67
4	0.51	0.53	0.85	0.00	0.00
5	0.49	0.46	0.68	0.00	0.00

Frequency Score (Monetary)

	1	2	3	4	5
1	406	1	0	0	0
2	1285	45	11	4	3
3	359	21	3	0	1
4	108	7	1	0	0
5	90	3	1	0	0

Decision Variables — Xij (Frequency Score, Monetary)

Xij	1	2	3	4	5
1	0%	100%	0%	0%	0%
2	8%	100%	100%	100%	100%
3	100%	100%	100%	0%	100%
4	100%	100%	100%	0%	0%
5	0%	0%	0%	0%	0%

Cost of Campaign (Frequency Score, Monetary)

	1	2	3	4	5
1	$0	$8	$0	$0	$0
2	$770	$338	$83	$30	$23
3	$2,693	$158	$23	$0	$8
4	$810	$53	$8	$0	$0
5	$0	$0	$0	$0	$0

Total Budget $5,000.00

Total Revenue (Frequency Score, Monetary)

	1	2	3	4	5
1	$0	$195	$0	$0	$0
2	$3,583	$10,518	$4,561	$2,382	$4,722
3	$24,857	$9,059	$1,389	$0	$1,473
4	$11,339	$4,142	$1,869	$0	$0
5	$0	$0	$0	$0	$0

$80,091

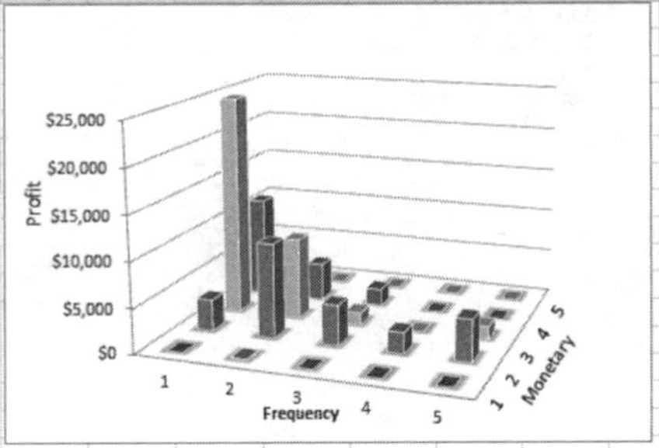

Figure 9-11 Solution for the continuous LP frequency-monetary model

LP Model with Three Dimensions

In this section, the LP model includes three variables of the RFM framework: *recency, frequency,* and *monetary value.* The objective remains the same—to maximize the expected revenue from potential customer purchases while not exceeding the budget constraints. The same OCR example is continued in this section and the *ch9_RFM_ LP_three_dimensions* file is used to illustrate this three-dimensional LP RFM model.

LP Model Formulation

Let the decision variable for this case be a 0-1 unknown variable as follows:

x_{ijk} = 1 if customers in recency i, frequency j, and monetary group k are reached; 0, otherwise

$$\text{Maximize: } Z_{rfm} = \sum_{i=1}^{R}\sum_{j=1}^{F}\sum_{k=1}^{M} N_{ijk}(p_{ijk}V_{ijk} - C)x_{ijk} \qquad (9.12)$$

subject to:

$$\sum_{i=1}^{R}\sum_{j=1}^{F}\sum_{k=1}^{M} N_{ijk}Cx_{ijk} \le B \qquad (9.13)$$

$$x_{ijk} = \{0,1\} \qquad i = 1...R \qquad j = 1...F \; k = 1...M \qquad (9.14)$$

where:

$i = 1...R$ index is used to identify the group of customers in a given recency category

$j = 1...F$ index is used to identify the group of customers in a given frequency category

$k = 1...M$ index is used to identify the group of customers in a given monetary category

N_{ijk} = number of customers in recency category i, frequency category j, and monetary category k

p_{ijk} = probability that a customer of recency category i, frequency j, and monetary k will respond to the campaign

V_{ijk} = average amount spent by a customer in recency category i, frequency category j, and monetary category k

C = average cost to reach a customer during the marketing campaign

B = available budget for the marketing campaign

Equation (9.12) is the objective function for the RFM LP model with three dimensions. It maximizes the expected revenue (Z_{rfm}) of the marketing campaign. As previously stated, a customer in a state of recency i, frequency j, and monetary k has a p_{ijk} chance of purchasing and $(1 - p_{ijk})$ chance of not purchasing. When purchasing, the expected revenue from a customer is calculated as $(V_{ijk} - C)$. When not purchasing, the expected revenue is simply $(-C)$. As such, the expected value of the expected revenue from a single customer in state ijk is:

$$p_{ijk}(V_{ijk} - C) + (1 - p_{ijk})(-C) \qquad (9.15)$$

which can be simplified as:

$$p_{ijk}V_{ijk} - C \qquad (9.16)$$

Because there are N_{ijk} customers with recency i, frequency j, and monetary k, the expected revenue from this group of customers is:

$$N_{ijk}(p_{ijk}V_{jk} - C) \qquad (9.17)$$

Equation (9.17) indicates the sum of expected revenues for all groups of customers for which a marketing decision to reach them $(x_{ijk} = 1)$ is made. Equation (9.13) ensures that the budget B for this marketing campaign is not exceeded. The left side of the equation represents the actual cost of the campaign, which is calculated as the sum of campaign costs for each group of customers in group ijk. Equation (9.14) represents the binary constraints for the decision variables x_{ijk}.

Solving the RFM Model with Three Dimensions

Because the customers are placed in three dimensional groups G_{ijk}, five separate worksheets are used to combine each monetary value (1, 2, 3, 4, and 5) with two dimensions (recency and frequency). Figure 9-12 shows the first worksheet (RF1), the combination of recency, frequency, with monetary value of 1. The number of customers in each group, average revenue for each group, and probability of return are calculated similarly to the cases with two dimensions. The only difference is that the COUNTIFS and AVERAGEIFS functions include three conditions. An IFERROR formula is also added here to consider combinations of recency, frequency, and monetary values that have no customers and to avoid the division by zero error. The following three formulas are used to calculate the values in cells C5 (revenue per customer in the group), C13 (probability that a customer in the group will purchase), and C21 (number of customers in the group):

- C5=IFERROR(AVERAGEIFS('RFM
  Summary'!$I$2:$I$2350, 'RFM Summary'!$E$2:$E$2350,
  'RF1'!C$4, 'RFM Summary'!$F$2:$F$2350, 'RF1'!$B5,
  'RFM Summary'!$G$2:$G$2350, 'RF1'!$H$1),0)

- C13=IFERROR(AVERAGEIFS('RFM
  Summary'!$H$2:$H$2350,'RFM
  Summary'!$E$2:$E$2350,C$12,'RFM
  Summary'!$F$2:$F$2350,$B13, 'RFM Summary'!$G$2:
  $G$2350,'RF1'!$H$1),0)

- C21=COUNTIFS('RFM
  Summary'!$E$2:$E$2350,C$20,'RFM
  Summary'!$F$2:$F$2350,$B21,'RFM
  Summary'!$G$2:$G$2350,$H$1)

In the preceding formulas, the RFM Summary worksheet holds the results of the pivot table and is used as source data for the previous parameter calculations. Note also the third condition in the previous formulas: The monetary range values ('RFM Summary'!$G$2:$G$2350) must be equal to $H$1 (which in our worksheet is holding the monetary value of 1). The formulas are then extended to the respective ranges of C4:G9, C13: G17, and C21:G25.

	A	B	C	D	E	F	G
3					Recency Score		
4		Vij	1	2	3	4	5
5		1	$23	$57	$60	$79	$77
6	Frequency	2	$0	$0	$0	$0	$356
7		3	$0	$0	$0	$0	$0
8		4	$0	$0	$0	$0	$0
9		5	$0	$0	$0	$0	$0
10							
11					Recency Score		
12		Pij	1	2	3	4	5
13		1	0.52	0.39	0.56	0.49	0.53
14		2	0.00	0.00	0.00	0.00	0.57
15	Frequency	3	0.00	0.00	0.00	0.00	0.00
16		4	0.00	0.00	0.00	0.00	0.00
17		5	0.00	0.00	0.00	0.00	0.00
18							
19					Recency Score		
20			1	2	3	4	5
21		1	330	19	18	15	24
22		2	0	0	0	0	1
23	Frequency	3	0	0	0	0	0
24		4	0	0	0	0	0
25		5	0	0	0	0	0
26							
27							

Raw_data / RFM Summary / start / **RF1** / RF2 / RF3

Figure 9-12 Parameters for LP recency, frequency, and M = 1 model

Figure 9-13 shows the decision variables (all set initially to one), calculations of the costs, and calculations of the expected revenue for the campaign. As shown by the formula on the side, the cost of the campaign is calculated as a product of the number of customers in a certain group, the decision variable value, and the cost per customer. Also, the expected revenue is calculated by multiplying the number of customers to reach, average revenue per customer, probability that the customer will respond, and decision whether customer is reached (1) or not reached (0) and then by subtracting the total cost for this group of customers.

The decision variables section is copied from the I9:M13 area of the final worksheet named Overall (shown in Figures 9.14 and 9.15). Worksheets RF2, RF3, RF4, and RF5 are built similarly to RF1 with two major differences: First, the monetary value held in cell H1 changes respectively to 2, 3, 4, and 5; and second, the values of decision variables are copied from different areas of the Overall worksheet. Specifically,

- For M = 2, decision variables are copied from I14:M18.
- For M = 3, decision variables are copied from I19:M23.
- For M = 4, decision variables are copied from I24:M28.
- For M = 5, decision variables are copied from I29:M33.

			L	M	N	O	P	Q	R	S
Decision Variables		Xij			Recency Score					
		Xij	1	2	3	4	5			
		1	0	0	0	0	0	<---=Overall!I9:M13		
	Frequency	2	0	0	0	0	0			
		3	0	0	0	0	0			
		4	0	0	0	0	0			
		5	0	0	0	0	0			
					Recency Score					
Cost of Campaign			1	2	3	4	5			
		1	$0	$0	$0	$0	$0	<---=$E$2*G21*P5		
	Frequency	2	$0	$0	$0	$0	$0			
		3	$0	$0	$0	$0	$0			
		4	$0	$0	$0	$0	$0			
		5	$0	$0	$0	$0	$0			
						Total Budget	$0.00			
					Recency Score					
Total Profit			1	2	3	4	5			
		1	$0	$0	$0	$0	$0	<---=G5*G13*G21*P5-P13		
	Frequency	2	$0	$0	$0	$0	$0			
		3	$0	$0	$0	$0	$0			
		4	$0	$0	$0	$0	$0			
		5	$0	$0	$0	$0	$0			
							$0			

Figure 9-13 Decision variables, cost, and expected revenue in the RF1 worksheet

Figure 9-14 shows the overall template and Solver Parameters dialog box for the model. The goal is to maximize the expected revenue (C12). The total expected revenue is the sum of expected revenues (cell P26) in worksheets RF1, RF2, RF3, RF4, and RF5. As a result, C12 is calculated as the SUM (start:end!P26), where *start* and *end* are two blank worksheets that are placed before and after RF1, RF2, RF3, RF4, and RF5 worksheets. Decision variables for this model are placed in cells I9:M33 and are constrained to be binary. Note that in Figure 9-14 all decision variables are initially set to 1.

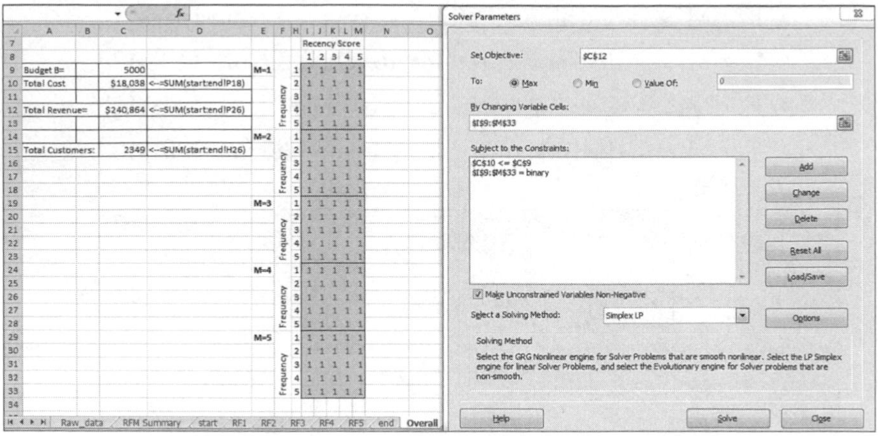

Figure 9-14 Template and Solver Parameters setup for the RFM LP model

The budget constraint C10 <= C9 indicates that the company should not spend more than $5,000. The total budget (cell C10) is calculated as the sum of budgets (cell P18) in worksheets RF1, RF2, RF3, RF4, and RF5. As a result:

$$C10 = SUM(start:end!P18)$$

The results of the Solver solution are shown in Figure 9-15. The solution indicates that the company should target the following groups of customers:

- For monetary value 1, the company should not contact any customers.

- For monetary value 2, the company should reach customers with recency 3 and frequency 2, recency 4 and frequency 2, recency 5 and frequencies 2, 3, 4, and 5.

- For monetary value 3, the company should contact customers with recency 3 and frequency 2. Also in this group, customers with recency 5 and frequencies 2 and 3 should be reached.

- For monetary value 4, only customers with recency 4 and frequencies 2 and 3 should be contacted.

- Finally, for monetary value 5, the company should contact customers with recency 4 and frequency 1, and customers with recency 5 and frequencies 1 and 2.

This solution is achieved under the budget constraint ($4,995) and maximizes the expected revenue ($230,493).

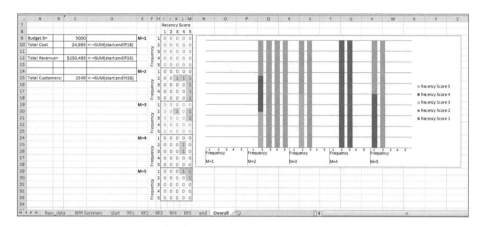

Figure 9-15 Solution for the three-dimensional LP RFM model

A Goal Programming Model for RFM

A marketing analyst wants to extend the previous LP models by adding priorities in each of the dimensions of the RFM approach. Assume that one-dimensional LP models are solved as described in Chapter 8 and the maximum expected revenue for each individual dimension is respectively P_R for recency, P_F for frequency, and P_M for monetary. Ideally, the decision maker wants to achieve all these goals; however, because it might not be possible to reach all of the goals simultaneously, the modeler will create a set of priorities for reaching a goal. These priorities depend on the importance of reaching a particular segment. For example, assume that the analyst values recency (R) more than frequency (F), and then F more than monetary value (M). Based on this assumption, the priorities could be P_R, P_F, and P_M respectively, where $P_R > P_F > P_M > 0$. The modeler will then create a new set of variables S_R, S_F, and S_M to represent the failure of meeting each priority and create a goal programming model. To illustrate the proposed GP model, the same sample of 2,349 customers from the online coffee retailer (OCR) is used. The *ch9_RFM_GP* file is used to illustrate the goal programming model to the RFM approach.

GP Model Notations

Let:

i = index for the group of customers in a given recency category $(i=1,...,5)$

j = index for the group of customers in a given frequency category $(j=1,...,5)$

k = index for the group of customers in a given monetary category $(k=1,...,5)$

V_i = expected revenue from a return customer of recency i

V_j = expected revenue from a return customer of frequency j

V_k = expected revenue from a return customer of monetary value k

p_i = the probability for a customer in recency group i to make a purchase

p_j = the probability for a customer in frequency group j to make a purchase

p_k = the probability for a customer in monetary value group k to make a purchase

N_i = number of customers in recency group i

N_j = number of customers in frequency group j

N_k = number of customers in monetary group k

C = average cost of reaching a customer during the marketing campaign

B = budget limitation for the marketing campaign

The cutoff points and the respective revenue, probability of return, and number of customers for each category are shown in Figure 9-16. The goal of the company is still to decide whether or not to reach customers in a certain category considering a limited marketing budget of $5,000. Three separate LP models, one for each category, are formulated and solved using the methodology demonstrated in the single RFM dimension sections. Then, each of these solutions is incorporated into the last goal programming model, which aims to identify the best customer categories to achieve several priority goals.

Recency Cutoffs	Vi		pi		Ni
1/1/2013	1	$49.02	0.50		1373
4/1/2013	2	$171.06	0.50		164
7/1/2013	3	$164.62	0.52		169
11/1/2013	4	$211.57	0.56		176
2/1/2014	5	$335.36	0.51		467
			Total:		2349
Frequency Cutoffs	Vj		pj		Nj
0	1	$101.86	0.50		2248
10	2	$687.12	0.50		77
20	3	$1,380.57	0.51		16
30	4	$1,325.23	0.46		4
40	5	$1,937.04	0.81		4
			Total:		2349
Monetary Cutoffs	Vk		pk		Nk
$0	1	$32.57	0.52		407
$25	2	$111.92	0.50		1348
$50	3	$203.20	0.50		384
$75	4	$293.81	0.51		116
$100	5	$433.88	0.49		94
			Total:		2349

Figure 9-16 Cutoff points, revenues, probabilities, and numbers of customers for each category

GP Model Formulation

The following is the formulation, which minimizes the penalties of not reaching the goals:

$$\text{Minimize } Z = P_R * S_R + P_F * S_F + P_M * S_M \tag{9.18}$$

subject to:

$$\sum_{i=1}^{R} N_i(p_iV_i - C)x_i + S_R = V_R \tag{9.19}$$

$$\sum_{j=1}^{F} N_j(p_jV_j - C)x_j + S_F = V_F \qquad (9.20)$$

$$\sum_{k=1}^{M} N_k(p_kV_k - C)x_k + S_M = V_M \qquad (9.21)$$

$$\sum_{i=1}^{R} N_iCx_i + \sum_{f=1}^{F} N_fCx_f + \sum_{k=1}^{M} N_kCx_k \le B \qquad (9.22)$$

$$x_i = 1\, i = 1 \dots R \qquad (9.23)$$

$$x_f = 1\, f = 1 \dots F \qquad (9.24)$$

$$x_k = 1\, k = 1 \dots M \qquad (9.25)$$

In the preceding formulation, (9.18) represents the objective function. Minimization of S_R has a priority over minimization of S_F because S_R has a larger contribution coefficent ($P_R > P_F$). Similarly, minimizing S_M has the lowest priority. (9.19), (9.20), and (9.21) represent the new set of constraints added to the model to make sure that expected revenue goals, achieved by solving each of the one-dimensional LP models, V_R = \$95,295, V_F = \$45,883, and V_M = \$72,157, are set to be achieved. (9.22) ensures that the overall budget (B = \$5,000) is not exceeded. Finally, (9.23), (9.24), and (9.25) ensure that decision variables are binary values.

Solving the GP RFM Model

Figure 9-17 shows the initial template for the goal programming approach. As seen, the model tends to become complex, but the following points can be helpful in understanding it:

- The shaded area in the upper-left corner of the worksheet holds the model variables. The model variables include deviation variables (C3:E3) and decision variables (C5:G7). The initial values for deviational variables are set to zero and initial values for decision variables are set to one.

- The values for revenues, probability of return, and number of customers for each subgroup are stored in the area M1:R14. These values are first calculated in the worksheets R, F, and M, which are used to solve the respective single-dimension LP model and then copied in the M1:R14 area.

- The values for revenues, probability of return, and number of customers from M1:R14 are then used in combination with the decision variables to calculate the campaign costs and total expected revenue stored in the B9:G16 range.

- The total expected revenue is summed in the cells H14:H16 and is carried out in the cells H5:H7, where it is compared against the optimal (maximum) expected revenues of each LP model respectively (stored in cells I5:I7).

Figure 9-18 shows Solver parameters for the model. The model seeks to minimize the objective function (G2), which is calculated as the product of weights (priorities) with the deviation variables ($= C2 \times C3 + D2 \times D3 + E2 \times E3$). The priorities in our application are assumed to be $P_R = 3$, $P_F = 2$, and $P_M = 1$. The By Changing the Variable Cells field contains the decision variables. The problem has four sets of constraints: (1) deviational variables must be positive, (2) decision variables must be binary, (3) total expected revenue + respective deviational variables must be equal to the optimal expected revenue of the previously solved LP models, and (4) the total budget should not exceed the available budget of \$15,000 (K8 <= K1 is a system constraint).

Figure 9-19 shows the final solution of the model as generated by Solver. As shown, the company should reach customers in recency group 5, frequency groups 2, 3, and 5, and monetary value group 5. This solution provides the highest priority to recency customers and the least priority to the monetary value customers. The total expected revenue for this solution is \$143,358 and the solution uses \$4,935 from the available \$5,000 budget.

	Recency Priority	Frequency Priority	Monetary Value Priority		Goal	Total Revenue		Budget	$ 5,000.00
Weights	3	2	1		0	$ 483,473		C=	$ 7.50
Slack	0.00	0.00	0.00						Used Budget
Score	1	2	3	4	5	Optimal Solution	Max		
Recency (Xi)	1	1	1	1	1	$ 162,074	$ 95,295	$ 162,074	$ 17,618
Frequency (Xj)	1	1	1	1	1	$ 162,147	$ 45,884	$ 162,147	$ 17,618
Monetary (Xk)	1	1	1	1	1	$ 159,252	$ 72,157	$ 159,252	$ 17,618
									$ 52,853
Campaign Cost-R	$ 10,297.50	$ 1,230.00	$ 1,267.50	$ 1,320.00	$ 3,502.50	$ 17,617.50			
Campaign Cost-F	$ 16,860.00	$ 577.50	$ 120.00	$ 30.00	$ 30.00	$ 17,617.50			
Campaign Cost-M	$ 3,052.50	$ 10,110.00	$ 2,880.00	$ 870.00	$ 705.00	$ 17,617.50			
Revenue R	$ 33,368.95	$ 14,122.39	$ 14,464.81	$ 20,839.58	$ 79,278.24	$ 162,074			
F	$ 115,506.01	$ 26,701.58	$ 11,251.63	$ 2,411.91	$ 6,276.01	$ 162,147			
M	$ 6,850.06	$ 75,789.87	$ 39,297.94	$ 17,464.34	$ 19,850.12	$ 159,252			

	v1	v2	v3	v4	v5
R	$49	$171	$165	$212	$335
F	$102	$687	$1,381	$1,325	$1,937
M	$33	$112	$203	$294	$434
	p1	p2	p3	p4	p5
R	0.50	0.50	0.52	0.56	0.51
F	0.50	0.50	0.51	0.46	0.81
M	0.52	0.50	0.50	0.51	0.49
	N1	N2	N3	N4	N5
R	1373	164	169	176	467
F	2248	77	16	4	4
M	407	1348	384	116	94

Figure 9-17 The initial template for the GP RFM model

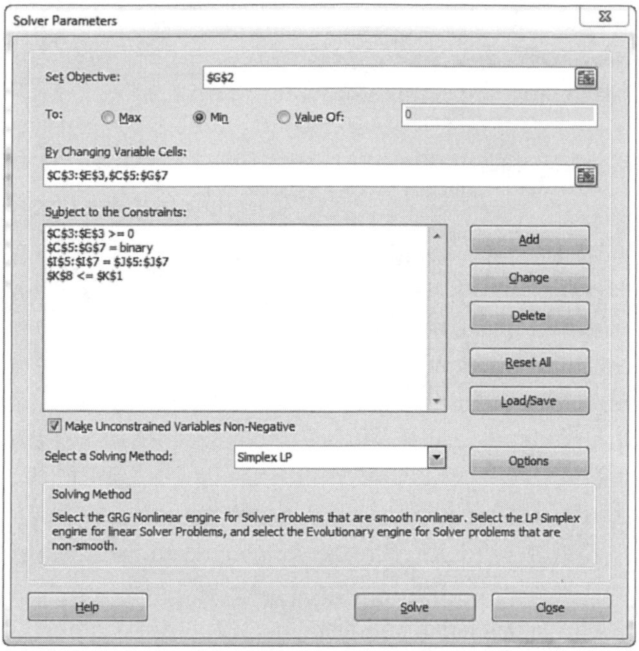

Figure 9-18 Solver Parameters setup for the GP RFM model

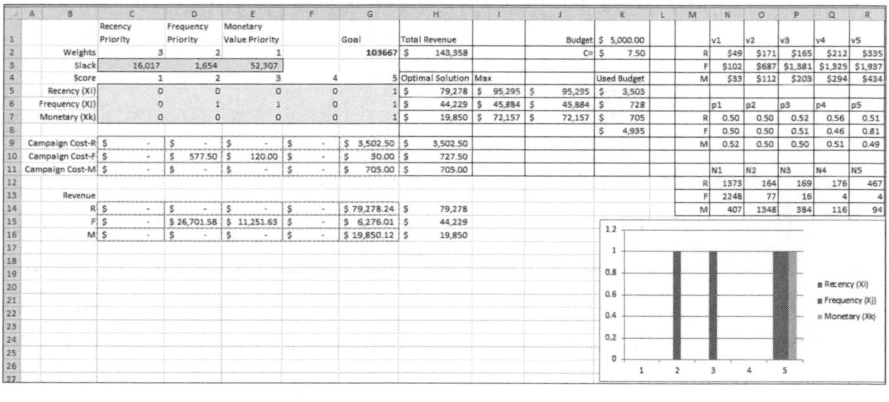

Figure 9-19 Final solution for the GP RFM model

The data analyst can explore possible scenarios by changing the priority values. For example, Figure 9-20 shows the final solution of the model with a different set of priorities: P_R = 30, P_F = 10, and P_M = 1. As shown, when giving the recency group more priority, the company should reach customers in recency groups 3 and 5, and frequency groups 3, 4, and 5. The total expected revenue for this solution drops to $113,683, and the campaign would use $4,950.

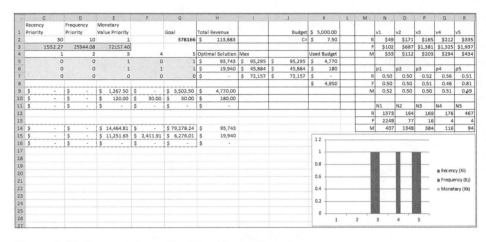

Figure 9-20 New solution with different priorities

Exploring Big Data with RFM Analytics

The critics of the RFM approach claim that this methodology is less likely to be used successfully in predictive and prescriptive analytics. "RFM analysis fails to say anything about the propensity of a prospect to respond to marketing stimuli. Rather, it simply shows who purchased from the company in the past."[67] The models proposed in the last two chapters bring the RFM approach into the era of Big Data: They augment the RFM analysis with predictive modeling, like customer response rates and prescriptive analytics, such as LP and GP models.

The combination of RFM analysis with predictive and prescriptive analytics in this text utilizes the RFM strengths and avoids its weaknesses. It is also an excellent example of combining intuition with analytics, a much-needed feature of today's business analytics. For example, for many companies, the majority of customers make only one purchase. That does not necessarily mean that the only segment produced by RFM will be the 1-1-1. A marketing analyst can intuitively assign different monetary values even for customers who have purchased previously. Further, the analyst can create the iso-value lines to segment customers into groups with similar value as a stepping stone toward incorporating more advanced mathematical programming approaches.

Wrap Up

The models discussed in this chapter allow the decision makers to establish RFM segments of customers who should be reached to maximize the expected revenues. Also, the models can help identify those RFM segments that are not worthy of pursuing because they are not profitable or because of marketing budget constraints. The problem of selecting target customer segments, given various priorities and the budget constraint, can also be treated as a multiobjective optimization problem with multiple obligatory and flexible goals with different priorities, interdependencies, and constraints on resources.

Review Questions

1. Among the three dimensions of the RFM approach, which one do you think is the most important: recency, frequency, or monetary value? Using a real-life business example, assign priorities and justify your ranking with business considerations.

2. Goal programming formulations in the chapter use cutoff points to determine different customer segments. What are some factors that the decision modeler must consider when setting these cutoff points?

3. Explain the logic of using the IFERROR function when calculating the input parameters of the LP models with two or three dimensions.

4. Describe a situation in which a data analyst may consider optimizing a marketing campaign with a special focus on recency and frequency only. Provide a business example to support your case.

5. Describe a situation in which a data analyst may consider optimizing a marketing campaign with a special focus on recency and monetary value only. Provide a business example to support your case.

6. Describe a situation in which a data analyst may consider optimizing a marketing campaign with a special focus on monetary value and frequency only. Provide a business example to support your case.

7. Describe a situation in which a data analyst may consider optimizing a marketing campaign with a high priority on recency. Provide a business example to support your case.

8. Describe a situation in which a data analyst may consider optimizing a marketing campaign with a high priority on frequency. Provide a business example to support your case.

9. Describe a situation in which a data analyst may consider optimizing a marketing campaign with a high priority on monetary value. Provide a business example to support your case.

10. Discuss the implications of combining an RFM approach with LP and GP models in the era of Big Data. List challenges and opportunities when the marketing analysis uses the RFM approach today.

Practice Problems

1. Consider the restaurant example described in problems 1 and 2 from Chapter 8. As a marketing analyst for the restaurant, you can use the same data set, the same cutoff points, and the same budget and cost requirements. However, you want to expand the analysis to include multiple dimensions in the same optimization model.

   a. Formulate and solve a 0-1 LP model that uses both recency and frequency data as input. The goal of the model is to identify those member segments that maximize the expected revenue under budget limitations. These segments are defined as a combination of recency and frequency values of members who should be included in the campaign.

b. Adjust the LP model so it identifies the percentage of customers from recency and frequency segments who should be included in the campaign.

c. Focus the analysis on the recency and monetary value of visits by members. Formulate and solve a 0-1 LP model to identify the recency and monetary value segments of members who should be included in the campaign. Adjust the model to identify the percentage of customers from each segment (recency and monetary value) to maximize the expected revenues under the same budget constraints.

d. Consider frequency and monetary value as the basis of your analysis. Formulate and solve a 0-1 LP model to identify the frequency and monetary value segments of members who should be included in the campaign to maximize the expected revenues under the same budget constraints. Also, adjust the model to identify the percentage of customers in each of the two segments.

e. Compare the results from the three two-dimensional binary LP models. Generate insights and recommendations for the restaurant management regarding the future of marketing campaigns and segments of customers who should be reached to maximize the expected revenues.

2. Continue the analysis of the restaurant from problem 1. You have the same budget constraints, but now you want to combine all three dimensions of the RFM analysis into a single optimization model.

a. Formulate and solve a 0-1 LP model that incorporates recency, frequency, and monetary value data as inputs. The goal of the model is to identify those combinations of segments that maximize the expected revenue under budget limitations. These segments are defined as a combination of recency, frequency, and monetary spending of members who should be included in the campaign.

  b. Formulate and solve a GP model considering the following priorities:

- Recency: Priority 1 (P1 = 1,000)
- Frequency: Priority 2 (P2 = 500)
- Monetary Value: Priority 2 (P2 = 500)
- Do Not Exceed Budget: Priority 3 (P3 = 100)

  c. Provide recommendations regarding the RFM segments that should be reached in the next marketing campaign. What is the expected revenue of your recommendation?

  d. Analyze the approaches and results between LP and GP models and discuss advantages and disadvantages of each approach.

  e. Change priorities listed in step b and observe the impact in the final solution. Overall, which of the three dimensions is most relevant to the marketing analysis in the restaurant?

3. Consider the shoe store example described in problems 3 and 4 from Chapter 8. Use the same data set, the same cutoff points, and the same budget and cost requirements.

  a. Formulate and solve a 0-1 LP model that uses both recency and frequency data as input. The goal of the model is to identify those combinations of recency and frequency values of customers who should be included in the campaign to maximize the expected revenue.

  b. Formulate and solve a continuous LP model so it identifies the percentage of customers from recency and frequency segments who should be included in the campaign to maximize the expected revenue under the same budget limitations.

  c. Focus the analysis on the recency and monetary value of purchases by customers. Formulate and solve a 0-1 LP model to identify the recency and monetary value segments of customers who should be included in the campaign.

  d. Adjust the model to identify the percentage of customers from each segment (recency and monetary value) to maximize the expected revenues under the same budget constraints.

 e. Now consider frequency and monetary value as the basis of your analysis. Formulate and solve a 0-1 LP model to identify the frequency and monetary value segments of customers who should be included in the campaign to maximize the expected revenues under the same budget constraints.

 f. Adjust the model to identify the percentage of customers in each of the two segments.

 g. Compare the results from the three two-dimensional binary LP models. Generate insights and recommendations for the shoe store management team regarding the future of marketing campaigns and segments of customers who should be reached to maximize the expected revenues within a limited budget.

4. Expand the analysis of the shoe store from problem 3. You have the same budget constraints, but now you want to combine all three dimensions of the RFM analysis into a single optimization model.

 a. Formulate and solve a 0-1 LP model that incorporates recency, frequency, and monetary value data as inputs. The goal of the model is to identify those combinations of segments that maximize the expected revenue under budget limitations. These segments are defined as a combination of recency, frequency, and monetary spending of customers who should be included in the campaign.

 b. Formulate and solve a GP model considering the following priorities:

  • Monetary Value: Priority 1 (P1 = 10)

  • Frequency: Priority 2 (P2 = 5)

  • Recency: Priority 3 (P3 = 2)

  Provide recommendations regarding the RFM segments that should be reached in the next marketing campaign. What is the expected revenue of your recommendation?

 c. Formulate and solve the same GP model with new priorities:

  • Budget Constraints: Priority 1 (P1 = 100)

  • Monetary Value: Priority 2 (P2 = 10)

  • Frequency and Recency: Priority 3 (P3 = 5)

Will your recommendations regarding the RFM segments that should be reached in the next marketing campaign change? Why or why not?

d. Overall, which of the above three dimensions is most relevant to the marketing analysis of the data from the shoe store?

5. Consider the home decorations example described in problems 5 and 6 from Chapter 8. As the marketing analyst for this store, you want to use a combination of RFM dimensions. Using the same data set, the same cutoff points, and the same budget and cost requirements from Chapter 8, perform the following analysis:

a. Formulate a 0-1 LP model that uses both recency and frequency data as input. The goal of the model is to identify those combinations of recency and frequency values of customers who should be included in the campaign to maximize the expected revenue. Solve the model and provide marketing recommendations based on your findings.

b. Formulate a continuous LP model to identify the percentage of customers from recency and frequency segments who should be included in the campaign to maximize the expected revenue under the same budget limitations. Solve the model and provide marketing recommendations based on your findings.

c. Focus the analysis on the recency and monetary value of purchases by customers. Formulate a 0-1 LP model to identify the recency and monetary value segments of customers who should be included in the campaign. Solve the model and provide marketing recommendations based on your findings.

d. Adjust the formulation to identify the percentage of customers from each segment (recency and monetary value) to maximize the expected revenues under the same budget constraints. Solve the model and provide marketing recommendations based on your findings.

e. Now consider frequency and monetary value as the basis of your analysis. Formulate a 0-1 LP model to identify the frequency and monetary value segments of customers who should be included in the campaign to maximize the expected revenues under the same budget constraints. Solve the model and provide marketing recommendations based on your findings.

f. Adjust the model to identify the percentage of customers in each of the previous two segments.

g. Compare the results from the three two-dimensional binary LP models. Generate insights and recommendations for the store management team regarding the future of marketing campaigns and segments of customers who should be reached to maximize the expected revenues within a limited budget.

6. Now, expand the analysis of the shoe store from problem 5. You have the same budget constraints, but now you want to combine all three dimensions of the RFM analysis into a single optimization model.

a. Formulate a 0-1 LP model that incorporates recency, frequency, and monetary value data as inputs. The goal of the model is to identify those combinations of segments that maximize the expected revenue under budget limitations. These segments are defined as a combination of recency, frequency, and monetary spending of customers who should be included in the campaign. Solve the model and provide marketing recommendations based on your findings.

b. Formulate and solve a GP model considering the following priorities:

- Frequency: Priority 1 (P1 = 500)
- Monetary: Priority 2 (P2 = 50)
- Recency: Priority 3 (P3 = 1)

Provide recommendations regarding the RFM segments that should be reached in the next marketing campaign. What is the expected revenue of your recommendation?

    c. Formulate and solve the same GP model with new priorities:
- Expected Revenue: Priority 1: (P1 = 1,000)
- Budget Constraints: Priority 2 (P2 = 100)
- Monetary Value: Priority 3 (P3 = 50)
- Frequency: Priority 4 (P4 = 5)

Will your recommendations regarding the RFM segments that should be reached change? Why or why not?

7. Consider the resort hotel example described in problems 7 and 8 from Chapter 8. As a marketing analyst for the hotel, you can use the same data set, the same cutoff points, and the same budget and cost requirements. Include multiple dimensions in your analysis.

    a. Formulate and solve a 0-1 LP model that uses both recency and frequency data as input. The goal of the model is to identify those hotel guest segments that maximize the expected revenue under budget limitations. These segments are defined as a combination of recency and frequency values of hotel guests who should be included in the campaign.

    b. Adjust the LP model so it identifies the percentage of hotel guests from recency and frequency segments who should be included in the campaign.

    c. Focus the analysis on the recency and monetary value of stays by hotel guests. Formulate and solve a 0-1 LP model to identify the recency and monetary value segments of hotel guests who should be included in the campaign. Adjust the model to identify the percentage of hotel guests from each segment (recency and monetary value) to maximize the expected revenues under the same budget constraints.

    d. Consider frequency and monetary value as the basis of your analysis. Formulate and solve a 0-1 LP model to identify the frequency and monetary value segments of hotel guests who should be included in the campaign to maximize the expected revenues under the same budget constraints. Also, adjust the model to identify the percentage of hotel guests in each of the two segments.

    e. Compare the results from the three two-dimensional binary LP models. Generate insights and recommendations for the hotel management regarding the future of marketing campaigns and segments of hotel guests who should be reached to maximize the expected revenues.

8. Continue the analysis of the hotel from problem 7. You have the same budget constraints, but now you want to combine all three dimensions of the RFM analysis into a single optimization model.

    a. Formulate and solve a 0-1 LP model that incorporates recency, frequency, and monetary value data as inputs. The goal of the model is to identify those combinations of segments that maximize the expected revenue under budget limitations. These segments are defined as a combination of recency, frequency, and monetary spending of hotel guests who should be included in the campaign.

    b. Formulate and solve a GP model considering the following priorities:

- Expected Revenues: Priority 1 (P1 = 1,000)
- Budget: Priority 2 (P2 = 500)
- Monetary Value: Priority 2 (P2 = 500)
- Recency and Frequency: Priority 3 (P3 = 100)

    c. Provide recommendations regarding the RFM segments that should be reached in the next marketing campaign. What is the expected revenue of your recommendation?

    d. Analyze the approaches and results between LP and GP models and discuss advantages and disadvantages of each approach.

10

Business Analytics with Simulation

Chapter Objectives

- Discuss the potential use of computer simulation to improve organizational performance
- Explore the role of simulation as a management science tool for optimization and decision making
- Discuss advantages and disadvantages of using simulation as a decision-making tool
- Provide examples of systems from real-world business situations and explain how simulation can be used to improve such systems
- Distinguish between discrete and continuous simulation models and their ability to replicate business settings
- Distinguish between static and dynamic simulation models and their ability to replicate business settings
- Distinguish between deterministic and stochastic simulation models and explore business situations where these models can be used
- Discuss the four basic elements of a computer simulation model: entities, locations, processes, and resources
- Suggest a simulation methodology that can be used to model business situations in the era of Big Data and underscore the importance of following each step in the methodology

- Discuss potential sources of data inputs for simulation models and how Big Data has changed the process of data collection
- Understand the concept of validation and verification as an important step in the simulation process

Prescriptive Analytics in Action: Blood Assurance Uses Simulation to Manage Platelet Inventory[1]

Blood Assurance is a full-service regional blood center serving more than 70 health-care facilities in Tennessee, Georgia, Alabama, Virginia, and North Carolina. Founded in 1972, Blood Assurance has 14 locations and 14 bloodmobiles to collect life-saving blood products used by area patients. One of the primary goals of the company is to meet the demand for platelets and minimize waste. Platelets are transfused to prevent or treat bleeding. Inventory management of platelets is complicated by unpredictable demand for a product with a shelf life of only a few days. In addition, health-care facilities in the area request platelets that match the blood types of patients identically (type-specific platelets).

The company requested assistance in developing a simulation-based decision support system to investigate, design, and test alternative strategies for platelet collection. The objective was to develop a platelet collection strategy that would reduce waste and meet demand for type-specific platelets. The simulation methodology allows modeling of complex and stochastic problems. The simulation-based decision making model was able to mimic the complexity of the blood inventory management system. Using historical records of previous donations and demand for platelets of each blood type, the simulation model was able to suggest appropriate collection strategies to reduce platelet waste by 50% and decrease unmet demand for type-specific platelets by 16%.

---

[1] This story is provided by Elizabeth Culler, MD, Medical Director at Blood Assurance.

Introduction

This chapter demonstrates the use of simulation methodology as a prescriptive analytics tool. Simulation is one of the most preferred techniques when investigating the behavior of complex business models. Simulation is now widely accepted as an effective method to assist management in evaluating different operational alternatives. Using simulation, a decision maker is able to mimic the complexity of the system and test alternative scenarios. The practicality of the simulation methodology is reduced by its inability to achieve optimal solutions. However, when used in a case-by-case basis, the simulation methodology can generate acceptable policies that are nearly optimal. When embedded in a decision support system with a user-friendly interface, the simulation methodology has the potential to become a powerful decision-making tool to create alternative scenarios and predict future revenues, costs, or risks for such scenarios.

Simulation offers several advantages over other mathematical programming techniques. From a practical perspective, a simulation model can be used to investigate a wide variety of "what if" questions about real-world systems before implementing potential costly changes. For example, different machines with different capabilities can be tested without actually purchasing and installing new equipment. Simulation can test new facility locations, product designs, or scheduling policies without any cost disruptions. New operating procedures and information flows can be explored without distracting ongoing operations in the real system. Designing and building simulation models often requires advanced training on simulation principles and simulation software.

Basic Simulation Terminology

A computer simulation model imitates the operations of a facility or process. These operations or business processes are treated as systems and are designed based on assumptions or approximations. If the structure of the systems is relatively simple, the modeler can use mathematical methods to answer questions of interest. When the systems are complex, simulation can be used to replicate the model, generate and collect data, and analyze the results. Because most business systems are complex, simulation models can become a powerful tool for decision makers. Understanding the simulation model starts with some basic terminology.

System

A *system* is a collection of entities and subentities that interact with each other as they process input into output. For example, a chemical production system is a collection of instruments, raw materials, and people who use chemical processes to transform raw materials into finished products. A school system is a collection of classrooms and instructors who process students through several courses to produce educated graduates. A restaurant system consists of several waiters, chefs, and tables, which receive customers, feed them, and process them through the restaurant.

A simulation model is also a system because it is used to represent a real-life system. The production facility, a chemical process, or the restaurant can be modeled into a computer model with entities (machines, workers, instructors, classrooms, or tables), input (order arrival quantities and times, processing times, number of machines, number of students enrolled in a class, employees working in a given shift, or number of waiters), or output (number of units assembled, percentage of defective products, happy diners, or student grades).

State of a System

As mentioned previously, a system includes several input, processing, or output variables. These variables hold values, which change during the simulation run. The state of the system is the set of variables necessary to describe the system and their values at a given point. For example, the number of waiters, number of customers being served, arrival time of the next party, and expected departure time of the next leaving customers can collectively describe the state of the restaurant at any point in time.

Discrete Versus Continuous Models

A simulation model is discrete when the state of the variables changes at discrete points in time. A restaurant, for example, is a discrete system, and can be represented with a discrete simulation model. The state variables of the restaurant change when a customer arrives or when an order is taken or when the food is served. A continuous simulation model, on the other hand, has variables whose state changes continuously. Changes in the continuous models are tracked over time according to a set of equations, typically involving differential equations. For example, a chemical process, a weather system, or airflow dynamics in new vehicles can all be simulated with continuous systems.

Static Versus Dynamic Simulation Models

A static simulation model represents the system at a given point in time. Static simulation models are sometimes referred to as Monte Carlo simulation models. For example, a marketing manager for a firm needs to estimate the annual net profit for a product, which depends on a certain variable, which is the fixed cost as well as on a series of uncertain variables, such as sales volume in units, price per unit, and unit costs. The manager can create a static simulation model that generates random values for the uncertain variables and calculates the estimated annual profit.

A dynamic model, however, represents the system over a period of time. Dynamic models have an embedded simulation clock, and the model is run from a given start time to a given end time. For example, daily weather models, restaurant shifts, and virus propagation can be simulated with dynamic models.

Deterministic Versus Stochastic Simulation Model

A deterministic simulation model contains no random variables. For example, assume that a small business manager wants to predict the annual sales for the next year. The amount of sales depends on the economic outlook and inflation rate. Note that this "prediction" is simply a hypothetical what-if statement. What would be the maximum annual sales if the economic outlook is recession and the inflation is low? Stochastic simulation models, on the other hand, have at least one random input variable. Because one or more input variables are random, the output of stochastic models is also random. For example, a queuing model with random arrivals and random processing times is a stochastic model.

Simulation Methodology

This section offers a systematic approach that can be used by practitioners to successfully adopt simulation as a decision-making tool. Simulation modeling is a complex process requiring careful planning and implementation. The process of developing a simulation study includes multiple steps. These steps are shown in Figure 10-1 and are briefly explained in the following sections.

Problem Description

Every simulation study starts with a good understanding of the business setting, which is the basis of the simulation model. Requests for proposal documents, business reports, interviews with the management team, and other sources can be used by the modeler to better understand the business description. An important part of understanding the business domain is establishing the purpose and scope of the simulation model. Just as in the linear and other programming models, the purpose of the study can help establish the objective function or dependent variables. The scope of the study is defined by the (a) time required to complete the simulation study and (b) the organizational unit that is included in the study. It is suggested that the time to complete a simulation study may vary from several weeks (for smaller problems) to several months (for more complex problems). Simulation studies are normally limited to systems that encompass isolated business units. However, the study may often include more complex systems with many entities, resources, time horizon, and outside dependencies; it very often encompasses a whole organization.

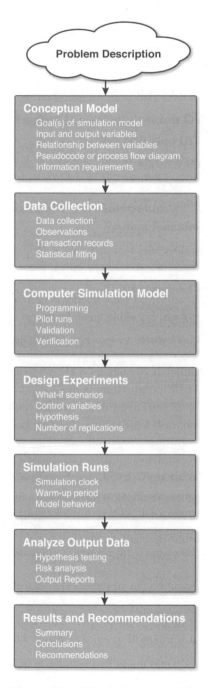

Figure 10-1 Simulation methodology

Conceptual Model

No matter the size of the system under investigation, the very first step for the decision maker is to design the model conceptually. This process starts with identifying the goals of the model. For example, the decision maker may want to determine the best inventory policy for reducing waste, the best factory layout for minimizing the internal transportation cost of moving components in the floor, or the optimal production lot for meeting customer demand while decreasing back-order levels.

Once the goals are established, the modeler identifies the set of input variables for the model. As mentioned earlier, some of these variables are stochastic (random) and some are deterministic. The type of decision variables defines the type of simulation model: stochastic or deterministic. For example, when simulating operations in a health-care center, patient arrival rate is an input variable. When patients arrive with appointments, it is relatively safe to assume that the arrival rate is not random. However, if the center is a walk-in clinic or an emergency room, then the arrival rate is random.

Further, relationships between variables are explored. Very often, intermediate variables are calculated, which later lead to output variables. The number of patients in the waiting room for a health-care center or the processing costs of components in a manufacturing cell are in-process variables. The number of patients served during the day for a health-care center, number of daily deposits in a bank, or number of products produced in a manufacturing cell are considered as output variables. When one or more input variables are random, then the output variables are random as well. At this stage, a high-level flow diagram or a pseudocode for future programming can be designed to better explain the logic of the future simulation model. The final goal of the conceptual model is to produce a list of information requirements. For this step, the modeler needs to answer the following questions:

- What information is needed to build the simulation model?
- What information is already available?
- What information needs to be collected?

The answer to the third question leads to the next step: data collection.

Data Collection

Two very important components of building the model are data collection and the generation of random distributions. Data is collected through historical records, system documentations, personal observations, and interviews. For example, when simulating a health-care center, data about patient flow time, patient arrival rates, time spent in the waiting room, blood tests, nursing station, and so forth can be recorded. When simulating a bank, patient arrivals can be observed and recorded; processing time at the teller window can also be recorded.

Before being used in the model, data should be tested for independence and homogeneity, and if stochastic, it must be fitted to statistical distributions. The decision maker must ensure that input variables are independent. For example, when a bank wants to measure the value of its customers, the modeler should consider two input variables: number of deposits during a month and the total amount of deposits for the month. These two measures, however, are not independent. The total amount of deposits often depends on the number of deposits. An independent set of variables would be the number of deposits and the amount of each deposit.

The decision maker must also ensure that input variables are homogenized. For example, continuing with the bank example, the number of deposits and the amount of each deposit may vary significantly by the type of account. The number of deposits during a month and the amount of each deposit can be significantly different for business accounts compared with individual accounts. So, these two measures must be counted separately for each account group.

Finally, the decision maker must represent the input variables as to their deterministic or stochastic values. Continuing with the bank example, the deposit amount can be entered in the model as a static value (for example, the average amount of each deposit) or as a stochastic value. In the latter case, a series of deposits can be statistically analyzed and a statistical distribution with estimated parameters is suggested. For example, the modeler can use a normal distribution to simulate the amount of deposits, an exponential distribution to model the frequency of the deposits, and a Poisson distribution to simulate the customer processing time during each deposit.

Computer Simulation Model

At this stage, a logical model of the process must be developed based on the data collected and the modeling constructs of the simulation software. Even though software packages have their own unique programming constructs, there is a great deal of similarity in the logical modeling. They normally include four basic elements, namely, (1) entities, (2) locations, (3) process flow for entities, and (4) resources. Further, several software packages provide the ability to generate random values, such as time for a given entity to move from one station to the next or time to process an entity in a given location using one of several probability distributions.

Because the output of stochastic models is random, the modeler must perform a pilot run with a limited number of replications. The results of the pilot runs must correspond with actual experiences derived in the real business settings. This allows the decision maker to generate enough solutions to perform model validation and verification. The validation process ensures that the simulation model represents the correct real-life system. The verification process ensures that the simulation model represents the real-life system correctly.

Design Experiments

One of the main advantages of simulation is the ability of the decision maker to conduct If-Then scenarios without actually making physical changes to the existing system. With a validated model, a variety of alternatives may be tested and optimized. The variables of the model can be viewed as decision variables and independent or control variables. If-Then analysis can be conducted using different values for decision variables. The best scenarios can be identified by stating the base and alternative hypothesis and testing this hypothesis in the next step. The minimum number of replications can be determined at this point using statistical formulas based on an acceptable level of error. Harrell, Bateman, Gogg, and Mott [68] provide an approach to computing the number of replications required to ascertain a selected degree of accuracy.

Simulation Runs

The model is run and hypothesis testing is performed. The simulation clock is also used in discrete simulation models as a variable that keeps the current value of simulated time in the model. Usually, there is no relation between simulated time and real time needed to run a model on a computer. A warm-up period is sometimes used to bring the system into a stable state, and after that the model parameters are calculated and the behavior of the model is observed. Many simulation software programs allow for a visual observation, and the decision maker can gain important insight by studying the changes of the state of the system over time.

Analyze Output

In this step, the statistical analysis of the output variables is conducted. Because input variables are random, the output measures are also random. While conducting the experiment, the minimum number of replications is performed and this allows for the generation of statistically significant data for reliable recommendations. Points and intervals are estimated to measure the performance of the system, and hypothesis testing and risk analysis are performed and output reports are prepared.

Results and Recommendations

The simulation methodology concludes with a summary of the results, main findings and conclusions, and, most important, practical recommendations. The best scenario is identified and the best combination of decision variables is recommended. Simulations results show what could happen under various scenarios—they do not show what will happen. Even with a very good simulation model, there are potentially many factors not included in the model that can determine the real-world outcome. Therefore, far-reaching decisions should not be based solely on the outcomes of simulations.

Simulation Methodology in Action

This section demonstrates the application of the simulation methodology in a simple Excel-based simulation model. The data used in this example is fictional, and the model is simply illustrative and can be used in other similar business situations. Any business that produces perishable products can use the same simulation methodology and template offered in the example. More complicated models can also be built with Excel using similar methodology. Very complicated models require complex Excel templates and specialized simulation software, such as ProModel, SAS Simulation Studio, Enterprise Dynamics, among others. You can find a comprehensive list of commercial and noncommercial simulation software at http://en.wikipedia.org/wiki/List_of_discrete_event_simulation_software.

Problem Description

A fictional Blood Bank Agency (BBA) wants to determine the optimal level of collection is in order to maximize the agency's revenue and meet the weekly demand for blood platelets. Although donations are made on a voluntary basis, assume that the agency spends about $150 to process collected blood units into one unit of blood platelets. To recover any costs, the agency charges receiving hospitals about $400 per platelet. There is also a $20 disposal cost for each unit of unused platelets.

Conceptual Model

Ideally, the agency should collect enough platelets to meet the demand but not exceed it. However, the platelet inventory management is complicated by unpredictable demand for a product with a shelf life of only a few days. If the agency collects more than the demand, then the difference must be discarded. If the agency collects less than the demand, then the agency cannot meet its contractual agreements with surrounding health-care centers.

Data Collection

The simulation model implements a stochastic pull system. Data used in the model can be retrieved from the operational activity of the blood center during one full year. The weekly demand for platelets is an uncontrolled variable and is governed by the discrete random distribution shown in Table 10-1.

Table 10-1 Distribution of Weekly Demand for Platelets

Weekly Demand for Platelets	Probability
300	.10
500	.25
800	.35
1000	.30

Computer Simulation Model

Figure 10-2 shows the main template used to simulate a random demand and to calculate the net revenue as demand varies. For now, the collection level is set to 300 platelet units per week (cell C2). Cell C3 hold a random value between 0 and 1, which is generated with the RAND() function in Excel.

The random demand is placed in cell C4. Notice that this cell uses a VLOOKUP function, which looks for the random value in cell C2, compares it with the cumulative distribution in column G2:G5, and returns the values in the second column in the range G2:H5. Random numbers greater than or equal to 0 and less than .10 will yield a demand of 300 platelets; random numbers greater than or equal to .10 and less than .35 will yield a demand of 500 platelets; random numbers greater than or equal to .35 and less than .70 will yield a demand of 800 platelets; and random numbers greater than or equal to .70 will yield a demand of 1,000 platelets.

	A	B	C	D	E	F	G	H	I
1		Calculation Table					Cumulative Dist	Weekly D	Probability
2		Collection Level	300				0	300	10%
3		Random Number	0.695434629	<---rand()			10%	500	25%
4		Demand	800	=VLOOKUP(C3, $G$2:$H$5, 2)			35%	800	35%
5							70%	1000	30%
6		Processing cost per unit	$150				above-=G4+I4		100%
7		Revenue per unit	$400						
8		Disposal cost per unit	$20						
9									
10		Total revenue	$120,000	<---=C7*MIN(C2, C4)					
11		Total processing cost	$45,000	<---=C6*C2					
12		Total disposal cost	$0	<---=IF(C4>C2, 0, (C2-C4)*C8)					
13		Net revenue	$75,000	<---=C10-C11-C12					

Figure 10-2 An Excel template for the simulation model

The total revenue in cell C10 is calculated as the product of revenue per unit (C7) and the minimum value between collection level (C2) and demand (C4). In other words, if the demand is more than the collection, then the revenue is based on the collection level; otherwise, the revenue is based on the demand level. The total processing cost is simply the product of processing cost per unit (C6) and the collection level (C2). The total disposal cost (cell C12) is zero when the demand is more than the collection level (C4 > C2), or the product of overcollection (C2 – C4) with the disposal cost per unit (C8). Finally, the net revenue is calculated as total revenue minus collection cost minus disposal cost (C10 – C11 – C12).

Design Experiments and Simulation Runs

The collection level is the decision variable in the model, and four different scenarios for different collection levels are created and shown in Figure 10-3. Each scenario assumes a collection level of 400, 600, 800, and 1,000 units. When pressing the F9 key, Excel refreshes the random variable and a new demand is placed in the demand cell. The decision maker needs to generate a large number of demands and analyze the output. In this example, the minimum number of replications is set to 1,000. Data tables in Excel can be used to replicate the model 1,000 times and save the output of each replication for each scenario.

	A	B	C	D	E	F	G	H
1		Calculation Table					Cumulative Distr	Weekly D
2		Collection Level	300				0	300
3		Random Number	0.970937155	<---rand()			10%	500
4		Demand	1000	=VLOOKUP(C3, $G$2:$H$5, 2)			35%	800
5							70%	1000
6		Processing cost per unit	$150				above-=G4+I4	
7		Revenue per unit	$400					
8		Disposal cost per unit	$20					
9								
10		Total revenue	$120,000	<---=C7*MIN(C2, C4)				
11		Total processing cost	$45,000	<---=C6*C2				
12		Total disposal cost	$0	<---=IF(C4>C2, 0, (C2-C4)*C8)				
13		Net revenue	$75,000	<---=C10-C11-C12				
14								
15	below-=C13							
16	$75,000.00		400	600	800	1000	Data Table	
17	1		$100,000	$108,000	$74,000	-$44,000		
18	2		$100,000	$24,000	-$10,000	$166,000	Row input cell: $C$2	
19	3		$100,000	$150,000	$200,000	$250,000	Column input cell: $J$8	
20	4		$100,000	$108,000	$200,000	$250,000		
21	5		$100,000	$24,000	$200,000	$40,000	OK Cancel	
22	6		$100,000	$150,000	$200,000	$166,000		
23	7		$100,000	$108,000	$200,000	$166,000		
24	8		$100,000	$24,000	$200,000	$250,000		
25	9		$100,000	$108,000	$74,000	$40,000		

Figure 10-3 Collection scenarios and net revenue for 16 (out of 1,000) data points

First, net revenue (C13) is transferred in cell A16, which is the upper-left cell in the data table. Then, values from 1 to 1,000 are placed in the range A17:A1016. The *Fill Series* from Home ->Editing in Excel can be used to automatically enter all 1,000 numbers in the A17:A1016 range. You can use the following steps to automatically fill the cells with numbers:

1. Enter a 1 in cell A17.
2. Select the cell and then, from the *Home->Editing ->Fill -> Series*.
3. In the *Series* dialog box, enter 1 as a *Step value* and 1,000 for and a *Stop value*.
4. Make sure that *Columns* is selected and then click OK.

The following steps can be used to simulate 1,000 iterations of platelet demand for each collection level:

1. Select the range (A16:E1016).
2. Click Table on the Data menu.
3. Select any blank cell as a column input cell (for example, C8).
4. Choose collection level (cell C2) as the row input cell.
5. Click OK to simulate 1,000 platelet demands for each collection level.

Analyze Output, Results, and Recommendations

Figure 10-4 summarizes the results for the net revenue for 1,000 replications for each collection level. The average and standard deviation formulas used as descriptive statistics indicate that the best collection level will be 800 platelet units per week. The confidence interval allows for a t-test comparison with alpha = 0.05. As shown, the scenario for the lower limit for 800 units is better than the upper-limit scenario at the 600 collection level (144,920.49 > 12,424.05). That means that the intervals of confidence for these two scenarios do not overlap, and as a result, the 800 collection strategy is significantly better than the strategy of collecting 600 platelets per week. However, when comparing the 800 unit collection scenario with the 1,000 collection scenario, there is no statistically significant difference (144,920.49 is not greater than 146,656.38). As a result, the recommendation is that the blood bank agency must establish a collection level goal between 800 to 1,000 unit platelets per week.

Collection level	400	600	800	1000	
Mean	95,968.00	127,026.00	149,558.00	$140,548	<----=AVERAGE(E17:E1016)
Standard Dev	12,379.02	38,691.09	74,823.36	98,554.86	<----=STDEV(E17:E1016)
Confidence Interval	767.25	2,398.05	4,637.51	6,108.38	<----=CONFIDENCE(0.05,K16,1000)
Upper Limit	96,735.25	129,424.05	154,195.51	146,656.38	<----=K15+K17
Lower Limt	95,200.75	124,627.95	144,920.49	134,439.62	<----=K15-K17

Figure 10-4 Summary of simulation results

Exploring Big Data with Simulation

Incorporating Big Data in simulation models is associated with both opportunities and challenges. Simulation models can use Big Data to provide more in-depth analysis and processing in advance. [69]

The volume of Big Data allows the simulation modeler to use larger data sets to better define statistical distributions, which then generate more reliable input for the simulation model. Simulation modeling focuses more on relationships and less on causality; as such, it can be an appropriate tool to deal with high complexity and large amounts of computations presented by Big Data.

The variety and velocity of Big Data, on the other hand, pose challenges for the simulation modeler. Simulation models use predefined entities, attributes, relationships, and their status over time. The changing nature of Big Data requires that simulation models must be adjusted often and validated before each run. For complex simulation models, this task is costly and time consuming. Finally, simulations show what could happen under various scenarios—they do not show what will happen. In the era of Big Data, there are potentially many factors not included in the simulation models that may determine the real-world outcome.

Wrap Up

This chapter discussed simulation modeling as a decision-making and optimization tool. The importance of simulation has increased significantly compared with other optimization methods in the era of Big Data. As velocity and variety of data increase, the decision makers turn to simulation as an appropriate tool for complex systems and uncertain data. As volume increases, decision makers can take advantage of statistical fitting software to better estimate statistical distributions of input variables. Simulation models allow the decision makers to compare alternative scenarios and control variables in ideal experimental conditions without altering the physical reality and, in so doing, save costs.

This chapter detailed a simulation methodology that can be used by practitioners. The use of Microsoft Excel is suggested for relatively simple models. The proposed methodology can also be used when specialized simulation software is utilized to model more advanced and complex business systems. There are several disadvantages and pitfalls when using simulations. The modeler should consider that stochastic simulations produce estimates that are often associated with noise. The model validation and verification processes described in the chapter are very important for the reliability of the simulation results. In addition, the practitioners must be aware that simulation models are generally expensive to develop and implement. Simulation expertise is generally rare, and simulation training and specialized simulation software is costly.

Simulations should always be accompanied by appropriate statistical analysis for both summary and analysis of large volumes of output generated by such models. The conceptual model proposed in the simulation methodology allows the decision maker to clearly identify the objective function up front. It also allows the decision maker to establish an appropriate level of detail and adequate design of simulation experiments.

Review Questions

1. Discuss the role of simulation as a management science tool for optimization and decision making. What are some advantages and disadvantages of using simulation as a decision-making tool?

2. What is a system? Provide examples of systems from real-world business situations and explain how simulation can be used to improve such systems.

3. What is a state of the systems? Use the same business situations from the previous questions and illustrate the state of such systems with examples.

4. What is the difference between discrete and continuous simulation models? Do you think business systems can be replicated better with discrete or continuous models? Explain why.

5. What is the difference between a static and a dynamic simulation model? Do you think business systems can be replicated better with static or dynamic models? Explain why.

6. What is the difference between deterministic and stochastic simulation models? Provide a business situation that can be modeled with a deterministic model. Provide a business situation that can be modeled with a stochastic model.

7. What is the importance of having the problem description as the very first step in the simulation methodology? What sources of information can be used to better understand the problem description of the simulation model?

8. What is the scope of the simulation study? How can the modeler define the scope?

9. What is a conceptual simulation model? Select a business system that can be simulated and provide examples of goals, input variables, relationships between them, or information requirements needed to build the simulation model.

10. Discuss potential sources of data inputs for simulation models. How have volume, variety, and velocity of Big Data changed the process of data collection, in particular, and simulation modeling, in general?

11. Discuss the concept of data homogeneity and its importance when simulating business systems.

12. What are the four basic elements of a computer simulation model? Provide examples of each using a business system as the domain of a simulation model.

13. What are pilot runs and why are they important for the simulation process? Explain validation and verification and the difference between them. Provide examples of each using a business system as the domain of a simulation model.

14. Explain what-if scenarios and discuss how simulation modeling can be used to make better decisions.

15. Discuss how simulation has changed in the era of Big Data. Also, discuss disadvantages and pitfalls when using simulations.

Practice Problems

1. You are the marketing manager for a soft drink company. You need to determine the amount of drink to produce next year and maximize total revenue. The company predicts the following economic situations and their respective probabilities: recession 25%, recovering 30%, stable 15%, and booming 30%. The production cost per drink is estimated to be $0.45. The demand for the drink in each type of economy is shown in the following table:

Economy	Probability	Demand
Recession	25%	20,000
Recovering	30%	35,000
Stable	15%	85,000
Booming	30%	100,000

Price per unit is determined by the actual demand and is shown in the following table:

Demand	Price/Unit
20,000	$0.70
35,000	$0.80
85,000	$0.90
100,000	$1.00

a. Create an Excel template that calculates the total revenue for various production levels.

b. Use a combination of RAND() and LOOKUP() functions to allow the template to randomly suggest production levels and price levels considering the distribution of the state of the economy.

c. Generate 1,000 data points and calculate the revenue for each of these data points.

d. Use summary statistics to decide the best production level that provides the highest-possible expected revenue.

2. A new airline wants to determine how many flights to assign between two cities. Each flight can carry 100 passengers and the demand is estimated to follow the distribution indicated in the following table:

Demand	Probability
100	0.05
200	0.2
300	0.4
400	0.3
500	0.05

It costs the airline $70 to fly a passenger and each empty seat is costing the airline $50. The airline is charging an average rate of $250 per seat for every ticket sold.

a. Create an Excel template that calculates the total revenue for a given number of flights between the two cities.

b. Modify the template with RAND() and LOOKUP() functions to allow for random generation of passenger demands.

c. Generate 500 data points and calculate the airline revenue for each of these data points.

d. Use summary statistics to decide the optimal number of flights between two cities that maximizes the expected revenue.

3. Consider the airline case from the previous problem and assume the following impact of the established price on attracting the existing customer demand for the two cities:

Price	Demand Attracted
Below $250	100%
$251–$300	80%
$301–$400	50%
Over $400	0%

When the price is above $400, passengers will use an alternative route between the two cities.

a. Modify the previously created Excel template to calculate the total revenue for a given number of flights between the two cities and for a given price tag.

b. Modify the template with RAND() and LOOKUP() functions to allow for random generation of passenger demands.

c. Create a matrix that calculates the average airline revenue for different numbers of flights and different price levels.

d. Use summary statistics to decide the optimal ticket prices and number of flights between two cities that maximizes the expected revenue.

4. You are the manager of a theater company and are preparing to launch the summer performance series in the Main Theater Hall. The Hall has a maximum 10,000 seats and the theater company pays $0.75 per seat. The size of the theater hall can be adjusted for smaller audiences, respectively for 8,000, 6,000, 4,000, or 2,000 seats. The production costs for each show is $4,500. The probability distribution for different number of sold tickets is shown in the following table:

Demand	Probability
1,500	10%
3,000	20%
5,000	25%
7,500	35%
9,000	10%

a. Create an Excel template that calculates the total revenue for a given number of tickets sold.

b. Modify the template with RAND() and LOOKUP() functions to allow for random generation of the number of tickets sold.

c. Generate 1,000 data points and calculate the theater revenue for each of these data points for each hall size option.

    d. Use summary statistics to decide the optimal hall size that maximizes the expected revenue.

5. Wireless Tower Properties (WTP) is a tower company that constructs towers that provide cellular telephone coverage in areas that lack adequate reception. Once a tower is constructed, carriers such as Verizon, AT&T, or T-Mobile can purchase the tower and "claim" the new covered area. A tower costs $25,000 to build and is leased to a carrier for an up-front price of $40,000 and monthly usage of $1,800. Towers are rented for a five-year period with a renewal option at the end of this period. The probability that any given number of towers can be rented is shown in the following table:

Demand	Probability
1	10%
2	25%
3	30%
4	25%
5	10%

WTP is running projections and wants to determine how many towers to build in order to generate the largest amount of profit. Use a five-year projection period to:

    a. Calculate the total revenue for any given number of towers built.

    b. Use 1,000 data points to calculate profit for each different number of towers built.

    c. Use summary statistics to decide the optimal number of towers to build that maximizes the expected profit.

6. A carpet-making manufacturer is considering purchasing a new machine. There are three options in the market, and the operational characteristics for each machine are shown in the following tables:

Machine	A
Setup time	0.75
Cleanup time	0.2
Machine cost	$500,000

Machine	B
Setup time	0.65
Cleanup time	0.4
Machine cost	$900,000

Machine	C
Setup time	1.25
Cleanup time	0.5
Machine cost	$550,000

Machine A costs $500k, but it consistently produces dashboards slower than the other two machines at a normal distribution rate with an average 45 minutes and standard deviation of 10 minutes per each carpet roll. Machine B is the most expensive, at $900k, but produces at a normal distribution rate with an average 15 minutes and standard deviation of 5 minutes per each carpet roll. Finally, Machine C cost $550k and produces at an average 35 minutes and standard deviation of 5 minutes per each carpet roll. The company works eight-hour shifts for 250 working days per year, and plans to sell the carpet units for $300 each.

a. Calculate the total revenue for any given machine to be purchased.

b. Generate 1,000 data points and calculate profit for each of these data points for a different machine.

c. Use summary statistics to select the best machine.

7. A local Chinese restaurant wants to determine the number of dishes to be prepared before operating hours for the most popular order, the Mongolian Beef. Mongolian Beef costs $15.95 to guests of the restaurant and costs the restaurant $7.00 to produce. Any dishes that are prepared, but not ordered, cost the restaurant $.50 per dish to dispose. Past data indicates that 300 dishes are sold 10% of the days, 400 dishes are sold 50% of the days, 600 dishes are sold 30% of the days, and 700 dishes are sold 10% of the days.

a. Create an Excel template that calculates the total revenue for any given number of dishes to be purchased.

b. Modify the template with RAND() and LOOKUP() functions to allow for random generation of number of dishes ordered.

c. Generate 1,000 data points and calculate profit for each of these data points for a different scenario.

d. Use summary statistics to select the optimal number of dishes to prepare.

A

Excel Tools for the Management Scientist

The following is a summary of the ten most used functions throughout the book. A brief description, syntax, and an illustration with an example are provided for each function. The *Appendix A* file contains sales records for a two-year period. You can download the *Appendix A* file from the companion website at www.informit.com/title/9780133760354. Each function discussed in this appendix is demonstrated in a separate worksheet in the file.

1: Shortcut Keys

In the era of Big Data, data scientists often deal with large data sets that spread over hundreds of columns and thousands or millions of rows. Shortcut keys offer an easy way to navigate through records, fields, and select a single value or range of values. Table A-1 shows the most popular shortcut key combinations when working with large data sets.

Table A-1 Summary of Shortcut Keys for Large Data Sets

Shortcut Key Combination	Navigation Result
Ctrl+down arrow	To get to the bottom row of your data set
Ctrl+up arrow	To get to the top row of your data set
Ctrl+right arrow	To get to the last column of your data set
Ctrl+left arrow	To get to the first column of your data set
Ctrl+Home	To get to the first cell (top left) in your data set
Ctrl+End	To get to the last cell (bottom right) in your data set
Any of the above +Shift	To select an entire range

Use the *shortcut keys* in Table A-1 to apply shortcut and perform the following tasks:

1. Quickly identify how many transactions are in the file.

   Start from any cell in the data set (for example, cell A2), hold the Ctrl key down, and press the down-arrow key. The bottom cell (A1081) is selected. Considering that the first two rows contain labels, there are 1,079 transactions in the data set.

2. What is the sales value of the last transaction in the record set?

   Sales are placed in the last column of the data set. Start from cell A1081 and press Ctrl+right arrow to move to cell D1081. The value for that cell is $1,023.18.

3. What is the sales value of the first transaction in the record set?

   Start from cell D1081 and press Ctrl+up arrow to move to cell D2. The sales value for the first record is $908.70.

4. Select all the dates of the transaction in the Transaction Sales column.

   Start from cell A3, hold the Ctrl and Shift keys down simultaneously, and press the down-arrow key. The range A3: A1081 is highlighted. Now, you can copy (Ctrl+C) or cut (Ctrl+X) the selected range.

5. Select all the transactions and all the columns for each transaction. Start from any cell in the data set and press Ctrl+A to select all rows and columns in the data set.

For a complete list of shortcut keys used in Excel, go to http://www.computerhope.com/shortcut/excel.htm.

For a complete list of general-purpose keyboard shortcut keys, go to http://www.computerhope.com/shortcut.htm.

	A	B	C	D	E	F	G	H	I	J	K
1		Example data									
2	Transaction date	Customer Name	Salesperson	Sales							
3	11/11/2014	Nathan	Lawrence	$908.70							
4	11/11/2014	Jonathan	Lawrence	$385.64		Salesperson	Sum If				
5	11/11/2014	Mathew	Lawrence	$535.79		Joseph	$202,392.79	=SUMIF($C$3:$C$1081,F5,$D$3:$D$1081)			
6	11/11/2014	Mathew	Lawrence	$456.67		Lawrence	$208,841.82				
7	11/11/2014	Mathew	Matt	$834.81		Matt	$231,051.99				
8	11/11/2014	Nathan	Matt	$502.92							
9	11/11/2014	Mathew	Matt	$302.40							
10	11/11/2014	Nathan	Matt	$482.00		Salesperson	Average If				
11	11/11/2014	Mathew	Matt	$423.15		Joseph	$593.53	=AVERAGEIF($C$3:$C$1081,F11,$D$3:G10)			
12	11/11/2014	Nathan	Matt	$719.45		Lawrence	$596.69				
13	11/11/2014	Mathew	Lawrence	$560.43		Matt	$595.49				
14	11/11/2014	Jonathan	Matt	$317.97							
15	11/11/2014	Mathew	Matt	$436.95							
16	11/11/2014	Mathew	Matt	$339.62		Salesperson	Count If				
17	11/11/2014	Jessica	Joseph	$849.22		Joseph	341	=COUNTIF($C$3:$C$1081,F17)			
18	11/11/2014	Jonathan	Lawrence	$455.54		Lawrence	350				
19	11/11/2014	Nathan	Joseph	$609.80		Matt	388				
20	11/11/2014	Jonathan	Lawrence	$736.94							
21	11/11/2014	Jonathan	Joseph	$582.45							
22	11/11/2014	Nathan	Lawrence	$548.37							

Figure A-1 Examples of using SUMIF, AVERAGEIF, and COUNTIF functions

2: SUMIF

The SUMIF function can be used to total values that meet specified criteria in a given range. For example, assume an analyst wants to determine the total sales for each salesperson. The SUMIF function can be used to add the value of sales that match each salesperson name. The syntax for the function is:

= SUMIF (range, criteria, [sum_range])

In the example, the above function is expressed as:

=SUMIF (range of salespersons, name of salesperson, range of sales)

Use the SUMIF-AVERAGEIF-COUNTIF sheet and apply the following formula, as shown in Figure A-1:

G5 = SUMIF ($C$3:$C$1081, F5, $D$3:$D$1081)

Extend the formula to cells G6 and G7.

3: AVERAGEIF

The AVERAGEIF function can be used to average the values that meet specified criteria in a given range. Now, assume the analyst wants to determine the average sales for each salesperson. The AVERAGEIF function can be used to calculate the average of the sales that match each salesperson name. The syntax for the function is:

= AVERAGEIF (range, criteria, [average_range])

In the example, the above function is expressed as:

=AVERAGEIF (range of salespersons, name of salesperson, range of sales)

Use the SUMIF-AVERAGEIF-COUNTIF worksheet in the *Appendix A* file and apply the following formula, as shown in Figure A-1:

G11 =AVERAGEIF ($C$3:$C$1081, F11, $D$3:G10)

Extend the formula to cells G12 and G13.

4: COUNTIF

The COUNTIF function can be used to count how many values meet specified criteria in a given range. For example, assume the analyst wants to determine the number of sales each salesperson has closed. In other words, how many times is the name of each salesperson listed in the Salesperson column? The COUNTIF function can be used for this purpose and the syntax is:

= COUNTIF (range, criteria)

In the example, the above function is expressed as:

=COUNTIF (range of salespersons, name of salesperson)

More specifically, using the example shown in Figure A-1, the formula as shown in Figure A-1, is:

G17=COUNTIF ($C$3:$C$1081, F17)

Apply this formula and extend it for cells G18 and G19.

5: IFERROR

If a formula cannot properly evaluate a result, Excel will generate an error value. These errors are usually noted by values such as #VALUE!, #NAME?, #DIV/0!, #NULL!, or #N/A. Often, the decision maker needs to ignore these errors or replace them with another value. For example, assume that in the transactions sales data set, the decision maker wants to calculate the average sales that each salesperson has generated by each customer. As shown in Figure A-2, the AVERAGEIFS[1] formula can be used for this purpose. For cell J3, the formula looks like:

$$J3 = AVERAGEIFS\ (\$D\$3{:}\$D\$1081,\ \$B\$3{:}\$B\$1081,$$
$$\$G3,\ \$C\$3{:}\$C\$1081,\ J\$2)$$

The result of this formula is $257.70. That means Andrew (condition 1) has purchased an average of $257.70 from Joseph (condition 2). However, when applying the same formula for combinations between Jessica and Lawrence, Jessica and Matt, and Mathew and Matt, #DIV/0! is the result. The reason for this result is that Jessica has never purchased from Lawrence or Matt, and Mathew has never purchased from Matt. So it will be better if the cell displays NA (not applicable).

The IFERROR function can be used in this situation. The IFERROR function tests a cell or calculation to determine whether an error has been generated. It will show TRUE for any type of error and FALSE if no error is found. Specifically, the syntax is:

$$=IFERROR\ (Cell\ to\ be\ tested)$$

where *Cell to be tested* can be a cell reference or a formula.

As shown in Figure A-2, replacing the previous formula with an IFERROR formula generates the correct results. For cell J14, the formula is:

$$=IFERROR\ (AVERAGEIFS\ (\$D\$3{:}\$D\$1081,$$
$$\$B\$3{:}\$B\$1081,\ \$G14,\ \$C\$3{:}\$C\$1081,\ J\$2),\ \text{“NA”})$$

---

[1] The AVERAGEIFS function is similar to AVERAGEIF, but can be used when there is more than one condition to be satisfied.

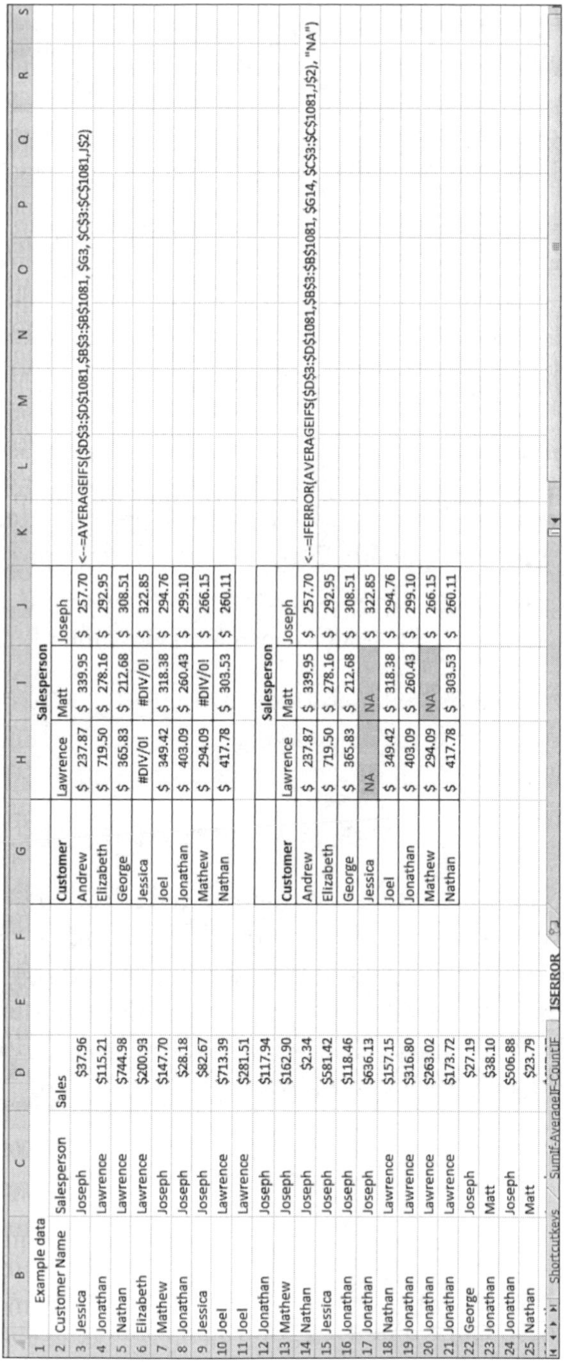

Example data

Customer Name	Salesperson	Sales
Jessica	Joseph	$37.96
Jonathan	Lawrence	$115.21
Nathan	Lawrence	$744.98
Elizabeth	Lawrence	$200.93
Mathew	Joseph	$147.70
Jonathan	Joseph	$28.18
Jessica	Joseph	$82.67
Joel	Lawrence	$713.39
Joel	Lawrence	$281.51
Jonathan	Joseph	$117.94
Mathew	Joseph	$162.90
Nathan	Joseph	$2.34
Jessica	Joseph	$581.42
Jonathan	Joseph	$118.46
Jonathan	Joseph	$636.13
Nathan	Lawrence	$157.15
Jonathan	Lawrence	$316.80
Jonathan	Lawrence	$263.02
Jonathan	Lawrence	$173.72
George	Joseph	$27.19
Jonathan	Matt	$38.10
Jonathan	Joseph	$506.88
Nathan	Matt	$23.79

Customer	Salesperson		
	Lawrence	Matt	Joseph
Andrew	$ 237.87	$ 339.95	$ 257.70
Elizabeth	$ 719.50	$ 278.16	$ 292.95
George	$ 365.83	$ 212.68	$ 308.51
Jessica	#DIV/0!	#DIV/0!	$ 322.85
Joel	$ 349.42	$ 318.38	$ 294.76
Jonathan	$ 403.09	$ 260.43	$ 299.10
Mathew	$ 294.09	#DIV/0!	$ 266.15
Nathan	$ 417.78	$ 303.53	$ 260.11

<--=AVERAGEIFS($D$3:$D$1081,$B$3:$B$1081, $G3, $C$3:$C$1081,J$2)

Customer	Salesperson		
	Lawrence	Matt	Joseph
Andrew	$ 237.87	$ 339.95	$ 257.70
Elizabeth	$ 719.50	$ 278.16	$ 292.95
George	$ 365.83	$ 212.68	$ 308.51
Jessica	NA	NA	$ 322.85
Joel	$ 349.42	$ 318.38	$ 294.76
Jonathan	$ 403.09	$ 260.43	$ 299.10
Mathew	$ 294.09	NA	$ 266.15
Nathan	$ 417.78	$ 303.53	$ 260.11

<--=IFERROR(AVERAGEIFS($D$3:$D$1081,$B$3:$B$1081, $G14, $C$3:$C$1081,J$2), "NA")

Shortcutkeys / SumIf-AverageIF-CountIF / ISERROR

Figure A-2 Illustration of IFERROR function

6: VLOOKUP or HLOOKUP

The vertical lookup (VLOOKUP) or horizontal lookup (HLOOKUP) functions can be used to search the first column of a range of cells, and then return a value from any cell on the same row of the range. The syntax for the VLOOKUP function is as follows:

=VLOOKUP(value, table_array, index_number, [not_exact_match])

where

- *value* is the value to search in the first column of the range of cells
- *table_array* is the range of cells
- *index_number* indicates the column where the return value exists (1 for first column, 2 for second column, and so on)
- *[not_exact_match]* can be TRUE if an approximate match is searched, or FALSE if an exact match is searched

Figure A-3 demonstrates the use of the VLOOKUP function to create an interactive display of selected cells. The function is placed in cell M17 as follows:

M17 =VLOOKUP (M15, $G$4:$J$12, M16, FALSE)

The function searches for the value placed in cell M15 (for example, Jessica), compares that value to the first column of table G4:J12 (list of customers), and returns the value found in the second column (for Lawrence). The result is NA. The use of VLOOKUP allows the decision maker to interactively display the average sales for each customer, generated by each salesperson.

G	H	I	J	K	L	M	N	O	P	Q	R
		Salesperson									
Customer	Lawrence	Matt	Joseph								
Andrew	$ 237.87	$ 339.95	$ 257.70								
Elizabeth	$ 719.50	$ 278.16	$ 292.95								
George	$ 365.83	$ 212.68	$ 308.51								
Jessica	NA	NA	$ 322.85								
Joel	$ 349.42	$ 318.38	$ 294.76								
Jonathan	$ 403.09	$ 260.43	$ 299.10								
Mathew	$ 294.09	NA	$ 266.15								
Nathan	$ 417.78	$ 303.53	$ 260.11								
				Enter Customer Name:		Jessica					
Enter Salesperson Column (Lawrence = 2, Matt = 3, Joseph = 4):						2					
				Average Sales:		NA	<--=VLOOKUP(M15, $G$4:$J$12, M16, FALSE)				

Figure A-3 Applying the VLOOKUP function

7: TRANSPOSE

The TRANSPOSE function copies data from a range, and places in it in a new range, turning it so that the data originally in columns is now in rows, and the data originally in rows is now in columns. The transpose range must be the same size as the original range. The function needs to be entered as an array formula.

The syntax of the function is as follows:

$$=\text{TRANSPOSE}(\text{Range})$$

The following steps can be followed to transpose rows into columns and columns into rows:

1. Select the range where the transposed table will be located.

   Make sure that the number of rows in the transposed table range is equal to the number of columns in the original table and the number of columns in the transposed table range is equal to the number of rows in the original table.

2. Enter **=Transpose (A2:D10)** in the upper-left corner of the selected, transposed table.

   In this example, *A2:D10* represents the range of the original table. Once the transposed table range is selected, simply click the = sign and enter the formula as shown in Figure A-4.

3. Simultaneously press the Ctrl+Shift+Enter keys and the values will be accordingly stored in the transposed table.

The results of the new table are shown in Figure A-5. If the values in the original table are changed, these changes are automatically reflected in the respective cells of the transposed table.

8: SUMPRODUCT

This function multiplies corresponding cells from two or more arrays and calculates the total of these products. The syntax of the function is as follows:

   SUMPRODUCT (array1, array2, array3...)

The arrays in the formula can be a continuous range of cells in a row or column and the number of cells must be the same in each array.

Assume that cells B2:B4 contain the average sale for each salesperson as calculated in a previous worksheet. Cells B8:B10 contain the number of sales transactions performed by each salesperson. To calculate the total sales, you multiply the number of sales by the average sales for each salesperson and then add these products. Figure A-6 indicates how this calculation is performed with the SUMPRODUCT function.

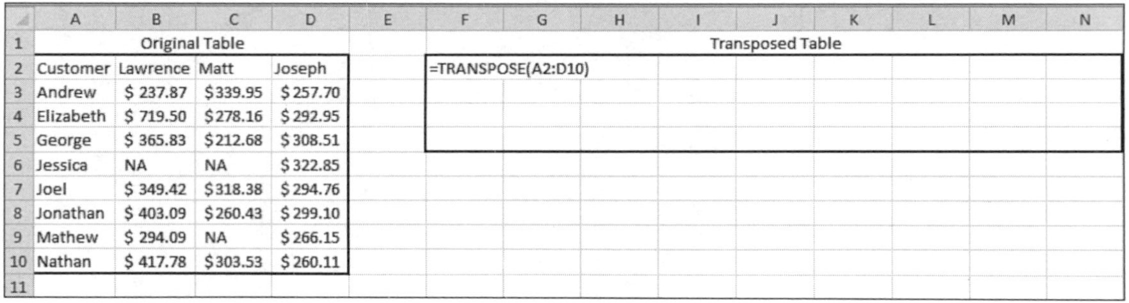

Original Table

Customer	Lawrence	Matt	Joseph
Andrew	$237.87	$339.95	$257.70
Elizabeth	$719.50	$278.16	$292.95
George	$365.83	$212.68	$308.51
Jessica	NA	NA	$322.85
Joel	$349.42	$318.38	$294.76
Jonathan	$403.09	$260.43	$299.10
Mathew	$294.09	NA	$266.15
Nathan	$417.78	$303.53	$260.11

Transposed Table

=TRANSPOSE(A2:D10)

Figure A-4 Selecting the range of the transposed table

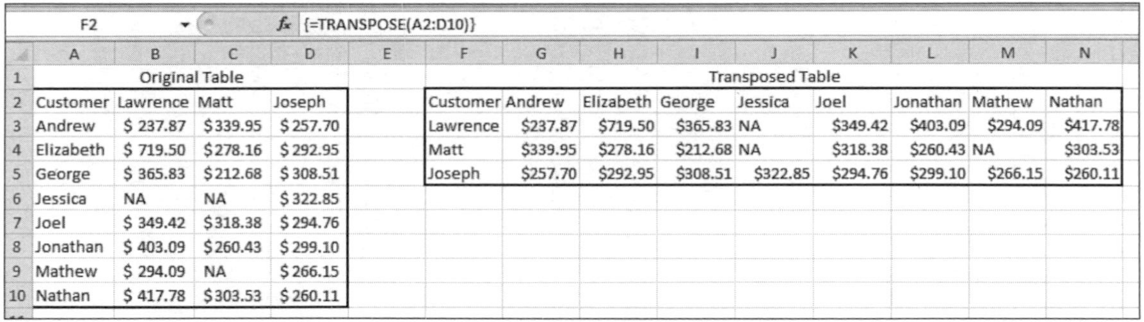

F2 · fx {=TRANSPOSE(A2:D10)}

Original Table

Customer	Lawrence	Matt	Joseph
Andrew	$237.87	$339.95	$257.70
Elizabeth	$719.50	$278.16	$292.95
George	$365.83	$212.68	$308.51
Jessica	NA	NA	$322.85
Joel	$349.42	$318.38	$294.76
Jonathan	$403.09	$260.43	$299.10
Mathew	$294.09	NA	$266.15
Nathan	$417.78	$303.53	$260.11

Transposed Table

Customer	Andrew	Elizabeth	George	Jessica	Joel	Jonathan	Mathew	Nathan
Lawrence	$237.87	$719.50	$365.83	NA	$349.42	$403.09	$294.09	$417.78
Matt	$339.95	$278.16	$212.68	NA	$318.38	$260.43	NA	$303.53
Joseph	$257.70	$292.95	$308.51	$322.85	$294.76	$299.10	$266.15	$260.11

Figure A-5 Transposed table: customers are displayed horizontally

	A	B	C	D	E
1	Salesperson	Average Sales			
2	Joseph	$290.94			
3	Lawrence	$402.47			
4	Matt	$281.51			
5					
6					
7	Salesperson	Number of Sales			
8	Joseph	678			
9	Lawrence	206			
10	Matt	195			
11					
12	Total Sales:	$335,060.65	<--=SUMPRODUCT(B2:B4,B8:B10)		

Figure A-6 Calculating total sales with SUMPRODUCT

The SUMPRODUCT function has performed the following calculations:

$$290.94 \times 678 + 402.47 \times 206 + 281.51 \times 195 = 335{,}060.65$$

Although the use of SUMPRODUCT seems trivial when there are only three cells in each array, you can see the potential advantage of using this function for larger arrays.

9: IF

The IF function can be used to return one value if a certain condition is TRUE and another value if the same condition is FALSE. The syntax of this function is: IF(logical_test, value_if_true, value_ if_false). The IF function has three components: logical test, value if true, and value if false.

Assume that the goal for each salesperson is to reach $100,000 in sales. The following IF statement can be used to assign a value next to each salesperson: IF (Sales>100,000, Achieved, Not Achieved). The upper section (range A1: C5) Figure A-7 shows the application of this function with a simple IF statement in Excel.

	A	B	C	D
1	Salesperson	Total Sales	Goal	
2	Joseph	$197,256.31	Achieved	<--=IF(B2>100000, "Achieved", "Not achieved")
3	Lawrence	$82,908.97	Not achieved	<--=IF(B3>100000, "Achieved", "Not achieved")
4	Matt	$54,895.38	Not achieved	<--=IF(B4>100000, "Achieved", "Not achieved")
5	Total:	$335,060.65		

Part a: Simple IF Statements

	A	B	C	D
7	Salesperson	Total Sales	Commission	
8	Joseph	$197,256.31	$29,588.45	<--=IF(B8<50000,B8*5%,IF(B8<80000,B8*7%,IF(B8<100000,B8*10%,B8*15%)))
9	Lawrence	$82,908.97	$8,290.90	<--=IF(B9<50000,B9*5%,IF(B9<80000,B9*7%,IF(B9<100000,B9*10%,B9*15%)))
10	Matt	$54,895.38	$3,842.68	<--=IF(B10<50000,B10*5%,IF(B10<80000,B10*7%,IF(B10<100000,B10*10%,B10*15%)))
11	Total:	$335,060.65		

Part b: Nested IF Statements

	A	B	C	D	E
13	Salesperson	Total Sales	Number of Sales	Commission	
14	Joseph	$197,256.31	678	$29,588.45	<--=IF(AND(B14<50000,C14>400), B14*5%,IF(AND(B14<100000,C14<400),B14*7%,IF(AND(B14<100000,C14>400),B14*10%,B14*15%)))
15	Lawrence	$82,908.97	206	$5,803.63	<--=IF(AND(B15<50000,C15<400),B15*5%,IF(AND(B15<100000,C15<400),B15*7%,IF(AND(B15<100000,C15>400),B15*10%,B15*15%)))
16	Matt	$54,895.38	195	$3,842.68	<--=IF(AND(B16<50000,C16<400),B16*5%,IF(AND(B16<100000,C16<400),B16*7%,IF(AND(B16<100000,C16>400),B16*10%,B16*15%)))
17	Total:	$335,060.65	1079		

Part c: Nested IF Statements with Multiple Conditions

Figure A-7 Applying IF statements to calculate sales commissions

Sometimes, the IF function can be used to evaluate more than two possible outcomes. In this situation, nested IF functions can be used. The nested statements still have three components (logical test, value if true, and value if false). However, a nested IF function replaces one or both 'if true/if false' components with another IF function.

Assume that you need to calculate the commission for each salesperson. The commission plan is based on the following criteria: If salespersons sell less than $50,000, they get 5% commission. If they sell between $50,000 and $80,000, they get 7% commission. If they sell between $80,000 and $100,000, they get 10% commission, and for sales over $100,000, they get 15% commission. Figure A-8 graphically displays the commission plan.

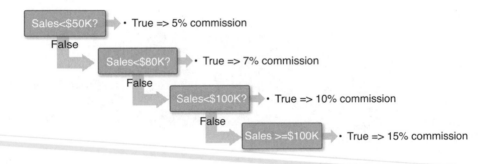

Figure A-8 Cutoff points for the commission plan

A nested IF formula can be used to implement the above plan. The formula used to calculate commission for Joseph is:

$$=IF (B8<50000, B8 \times 5\%, IF (B8<80000, B8 \times 7\%,$$
$$IF (B8<100000, B8 \times 10\%, B8 \times 15\%)))$$

The calculated commission for each salesperson and the application of nested IF statements are shown in the range A7: C11 of Figure A-7.

IF statements may also have multiple conditions in their first component. For example, assume that the commission plan is modified as follows:

- If salespersons sell less than $50,000 and make less than 400 sales, they get 5% commission.

- If they sell between $50,000 and $100,000 and make less than 400 sales, they get 7% commission.
- If they sell between $50,000 and $100,000 and make more than 400 sales, they get 10% commission.
- Otherwise, they receive 15% of sales in commission.

Figure A-9 graphically displays the new commission plan.

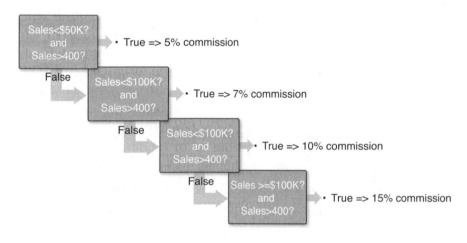

Figure A-9 New criteria for the commission plan

A nested IF formula with compound conditions can be used to implement the new commission plan. For Joseph, the formula is:

=IF (and (B14<50000, C14<400), B14 × 5%, IF (and (B14<100000, C14<400), B14 × 7%, IF (and (B14<100000, C14>400), B14 × 10%, B8 × 15%)))

The calculated commission for each salesperson and the application of nested IF statements are shown in the range A13: D17 of Figure A-7.

10: Pivot Table

Pivot tables are a powerful data summary tool. Pivot tables can be used to conveniently summarize large transactional data into averages, sums, maximum values, minimum values, and so on. Pivot table reports can also be used to show subtotals, custom formats, dynamic pivot charts, filtering, and sorting, and they can be used to drill down data.

The following steps can be used to create a pivot table:

1. Select the data.
2. Go to Insert and click on the Pivot Table option.
3. Select the target cell where you want to place the pivot table.
4. Create the pivot report with required criteria.

Assume that the analyst wants to summarize the 1,079 transactions in the *Appendix A* Excel file. Specifically, the pivot table must display the list of salespersons, total sales, average sales, and number of sales for each salesperson. Additionally, the analyst wants to display a list of customers and their total sales, average sales, and number of sales. These two reports can be easily created with two pivot tables. The steps for the first pivot table are shown in the following sections.

Step 1: Select the Data

All transactions must be included in the pivot table report, so select the range A2:D1081. As a reminder, you can use the shortcut key combination Ctrl+Shift+Down Arrow to reach the last transaction.

Step 2: Go to Insert and Click on New Pivot Table Option

The Create PivotTable Wizard is invoked and several choices are made. The Select a Table or Range section already shows the range of cells (A2:D1081) selected in the previous step. Often, the source of the data may reside in another computer and Excel allows you to select other sources by choosing the Use an External Data Source option. In addition, once the data source is selected, data can be refreshed (Options, Refresh) so the latest transactions are always included in the data analysis.

Step 3: Select the Target Cell Where You Want to Place the Pivot Table

The Create PivotTable window (Figure A-10) also allows you to select the New Worksheet or Existing Worksheet option. The New Worksheet option places the pivot table in a newly created worksheet. For this example, the Existing Worksheet option is selected.

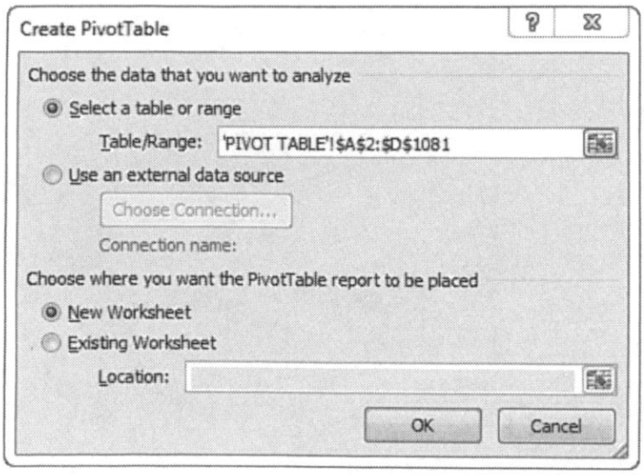

Create PivotTable	?	✕

Choose the data that you want to analyze

◉ Select a table or range

    Table/Range: `'PIVOT TABLE'!$A$2:$D$1081`

◯ Use an external data source

    [Choose Connection...]

    Connection name:

Choose where you want the PivotTable report to be placed

◉ New Worksheet

◯ Existing Worksheet

    Location:

    [OK] [Cancel]

Figure A-10 Create PivotTable Wizard

Step 4: Create the Pivot Report with Required Criteria

Figure A-11 shows the design portion of the Pivot Table Report. The pivot report is divided in to header and body sections. You can drag and drop the fields you want in each area. The body itself contains three parts: rows, columns, and cells. In the example, the Salesperson field is dragged and dropped in the Raw Labels area, and the Sales field is dropped three times in the Values area. The first time, the *sum* of sales is automatically calculated. The second time, the *average* is selected from the 'Value Field Settings' field, which can be accessed by clicking on the 'Average of Sales' field of the Values section of the 'PivotTable Field' list. At this time, you can select Value Field Settings—and choose Average from the drop-down menu. Similarly, Count of Sales is selected for the third sales field.

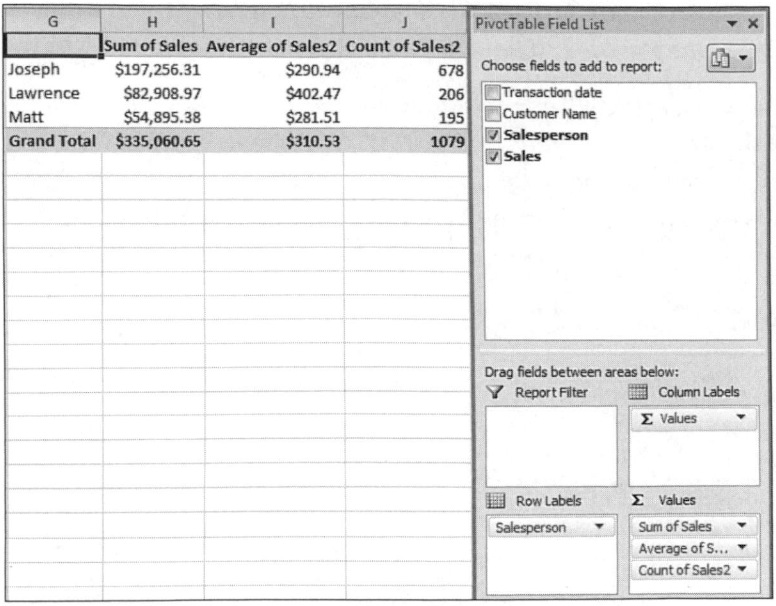

G	H	I	J	PivotTable Field List
	Sum of Sales	Average of Sales2	Count of Sales2	Choose fields to add to report:
Joseph	$197,256.31	$290.94	678	☐ Transaction date
Lawrence	$82,908.97	$402.47	206	☐ Customer Name
Matt	$54,895.38	$281.51	195	☑ Salesperson
Grand Total	$335,060.65	$310.53	1079	☑ Sales

Drag fields between areas below:

Report Filter / Column Labels / Σ Values

Row Labels / Σ Values

Row Labels: Salesperson

Values: Sum of Sales / Average of S... / Count of Sales2

Figure A-11 Pivot Table Report design

The same steps can be followed to display the summary of records for each customer. Figure A-12 shows the results of the two pivot tables.

	Transaction date	Customer Name	Salesperson	Sales			Row Labels ▾	Sum of Sales	Average of Sales2	Count of Sales2
2	Transaction date	Customer Name	Salesperson	Sales			Row Labels ▾	Sum of Sales	Average of Sales2	Count of Sales2
3	11/10/2014	Jessica	Joseph	$37.96			Joseph	$197,256.31	$290.94	678
4	11/10/2014	Jonathan	Lawrence	$115.21			Lawrence	$82,908.97	$402.47	206
5	11/8/2014	Nathan	Lawrence	$744.98			Matt	$54,895.38	$281.51	195
6	11/6/2014	Elizabeth	Lawrence	$200.93			Grand Total	$335,060.65	$310.53	1079
7	11/6/2014	Mathew	Joseph	$147.70						
8	11/5/2014	Jonathan	Joseph	$28.18						
9	11/4/2014	Jessica	Joseph	$82.67						
10	11/4/2014	Joel	Lawrence	$713.39						
11	11/1/2014	Joel	Lawrence	$281.51			Row Labels ▾	Sum of Sales	Average of Sales2	Count of Sales2
12	11/1/2014	Jonathan	Joseph	$117.94			Andrew	$4,848.60	$269.37	18
13	10/30/2014	Mathew	Joseph	$162.90			Elizabeth	$4,984.48	$383.42	13
14	10/30/2014	Nathan	Joseph	$2.34			George	$7,983.18	$295.67	27
15	10/29/2014	Jessica	Joseph	$581.42			Jessica	$78,451.51	$322.85	243
16	10/29/2014	Jonathan	Joseph	$118.46			Joel	$8,663.12	$320.86	27
17	10/27/2014	Jonathan	Joseph	$636.13			Jonathan	$82,705.64	$321.81	257
18	10/27/2014	Nathan	Lawrence	$157.15			Mathew	$64,814.45	$266.73	243
19	10/25/2014	Jonathan	Lawrence	$316.80			Nathan	$82,609.70	$329.12	251
20	10/24/2014	Jonathan	Lawrence	$263.02			Grand Total	$335,060.65	$310.53	1079
21	10/23/2014	Jonathan	Lawrence	$173.72						
22	10/22/2014	George	Joseph	$27.19						
23	10/19/2014	Jonathan	Matt	$38.10						
24	10/19/2014	Jonathan	Joseph	$506.88						
25	10/19/2014	Nathan	Matt	$23.79						
26	10/19/2014	Nathan	Joseph	$657.27						

Figure A-12 Summary of transactions with two pivot tables

B

A Brief Tour of Solver

Real-life problems are sometimes larger and more complex than those offered in this book. They might offer hundreds or even thousands of decision variables and constraints. However, the methodology is similar no matter the size of the problem. The following steps can be followed to solve LP models with Excel Solver:

1. Set up constraints and the objective function in Solver.
2. Select Solver options.
3. Generate the solution.
4. Analyze the results.

Setting Up Constraints and the Objective Function in Solver

Typically, Solver is not installed by default when Excel is first set up. To add Solver to the Tools menu, perform the following steps:

1. Select File -> Options -> Add-Ins.
2. In the Add-ins window, choose Manage: Excel Add-ins and click Go....
3. In the Add-ins dialog box, check Solver Add-In among the list of Add-ins available and Click OK.

Solver appears in the Analysis section of the Data tab. To access Solver, click the Data tab, and then click Solver in the Analysis section. The Solver Parameters dialog box shown in Figure B-1 opens. First, you select the objective cell. Solver will attempt to maximize or minimize the value of this cell. Sometimes, you might want to set the value of this cell to a certain level by selecting the Value Of option and entering the aspiration level in the respective box. Figure B-1 indicates that the target cell M2 is selected and is set to a Max. Next, enter the cell(s) containing the decision variables (in this example, $I$2:$I$49) into the By Changing Variable Cells field. Now, add the constraints by clicking the Add button. The Change Constraint dialog box shown in Figure B-2 opens.

The Cell Reference field normally holds the cells containing the left-hand side values, whereas the Constraint field contains the right-hand side values. You can use the drop-down menu in the center of the dialog box to select the following:

- <= when assigning a less than or equal to constraint
- = when assigning an equal to constraint
- >= when assigning a greater than or equal to constraint
- int when assigning integer values to the left-hand side cells
- bin when enforcing a binary {0, 1} value to the left-hand side cells
- dif when enforcing a different value for each decision variable

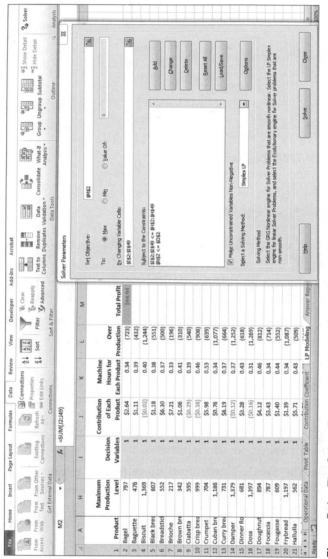

Figure B-1 The Solver Parameters dialog box

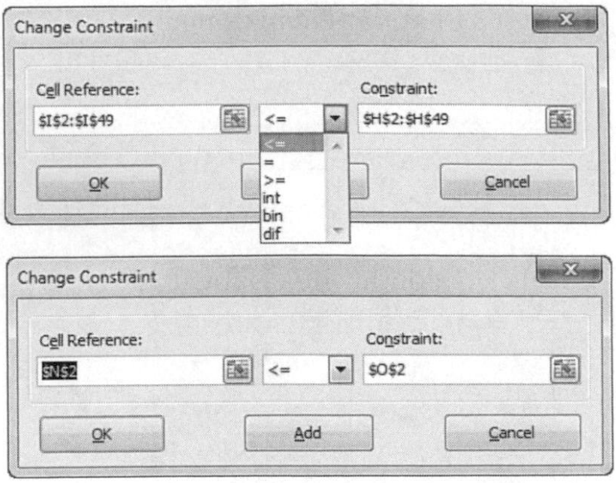

Figure B-2   Adding constraints with Solver

Selecting Solver Options

Solver offers several options that reflect various assumptions that the analyst can make regarding the LP model. The first option is to enforce that the solution values for the decision variables remain positive or zero. For example, when you check the Make Unconstrained Variables Non-Negative check box, the model will not return a negative number for the decision variables, as shown previously in Figure B-1.

The next option is to choose between different solution approaches—simplex LP, GRG nonlinear, and evolutionary:

- Select simplex linear programming when the objective function and constraints are linear functions.
- Select the GRG (Generalized Reduced Gradient) algorithm to solve problems when the problem is assumed to be nonlinear.
- Select the evolutionary method when Solver is trying to solve a nonsmooth problem.

Suppose you (the decision maker) have decided to produce either product M or product N. Or, assume that a given product starts to contribute only if it is produced in quantities more than X units per week. All these conditions make the behavior of the objective function or the behavior of selected constraints nonsmooth. In such cases, the evolutionary approach is suggested.

You can explore further options by clicking the Options button. Large models may require significant amounts of time and processing power to be executed and you must make a choice between a practical and good solution versus a nonpractical and optimal solution. These options allow you to choose between solution accuracy and practicality.

There are three tabs in the Options window: All Methods, GRG Nonlinear, and Evolutionary. Figure B-3 shows these options, and the following sections explain these options.

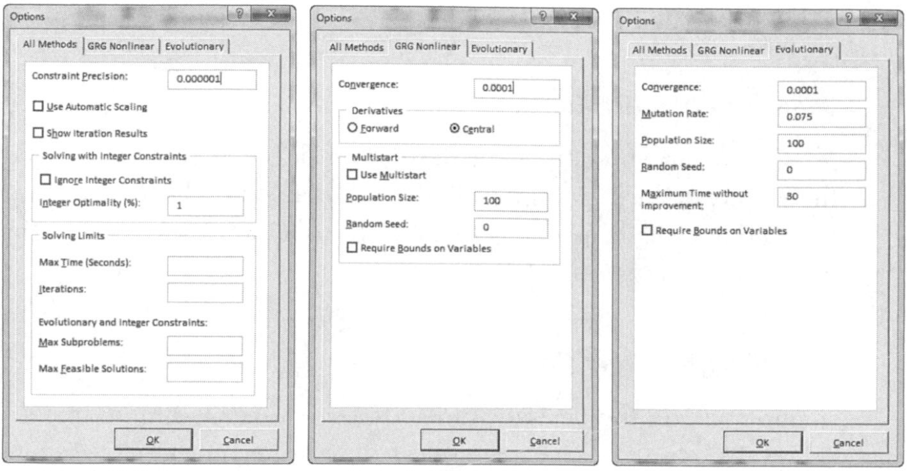

a. Options for All Methods   b. Options for GRG Nonlinear   c. Options for Evolutionary

Figure B-3   Further options in Solver

All Methods Options

You can use the first tab in the Options window to select general options applicable to all types of solution methods. The All Methods options include Constraint Precision, Use Automatic Scaling, Show Iteration Results, Ignore Integer Constraints, and other options to set up solution limits. You can accept the default values for these options unless you intentionally want to change the options, as discussed in the following sections.

Constraint Precision

In the Constraint Precision field, you choose the degree of precision. A specific degree indicates how much the relationship between the Cell Reference and the Constraint values can be violated. The smaller the number in the field, the higher the constraint precision.

Use Automatic Scaling

Check the Use Automatic Scaling check box if you want Solver to rescale the values of decision variables, constraints, and objective function to comparable magnitudes. This selection allows for a reduction of the impact of extremely large or small values on the accuracy of the solution process.

Show Iteration Results

Check the Show Iteration Results check box if you want to see the results of each attempt to find a solution. This check box is not checked by default, assuming that you (the decision maker) only want to see the final results.

Solving with Integer Constraints

Sometimes, the decision variables are constrained to be integer values. In those cases, you can select the Ignore Integer Constraints check box to relax such integer constraints. You can use the Integer Optimality % field to set the maximum percentage difference between the objective value of the suggested solution and the true optimal objective value. The default value is 1%, which means that Solver will stop anytime that the found solution is within 1% gap from the optimal solution.

Solving Limits

You can also choose the length of time, Max Time (Seconds), or the number of tries, Iterations, that Solver must run until it stops. For problems that include integer constraints on variables and for problems where an evolutionary approach is selected, you can impose further limits. Specifically, you can choose the maximum number of subproblems using the Max Subproblems field. Also, you can use the Max Feasible Solutions field to limit the maximum number of feasible solutions.

GRG Nonlinear Options

You can use the second tab in the Options window to select several options when the GRG nonlinear solving method is used. These options include Convergence, Derivatives, and multistart options for global optimization. Just as with the All Methods options, you can accept the default values for these options unless you intentionally want to change the options as discussed in the following sections.

Convergence

You might consider stopping Solver when there are no longer significant improvements in the value of the objective function as the iterations continue. In this situation, the value entered in the Convergence field indicates the amount of relative change to be allowed in the last five iterations before Solver displays the Solver Converged to the Current Solution message and stops further attempts.

Derivatives

The nonlinear method uses the first derivatives to approximate a solution during any given iteration. Mathematically, Solver uses two types of differencing to calculate the value of the derivative: forward differencing and central differencing. *Forward differencing* is calculated as:

$$f'(x) \cong \frac{f(x+h) - f(x)}{h}$$

while *central differencing* can be calculated as:

$$f'(x) \cong \frac{f(x+h) - f(x-h)}{2h}$$

Generally, *central differencing* yields more accurate derivatives, but this method requires more calculations. *Forward differencing* is the default value in this option.

Multistart

You can find a better solution by selecting the Use Multistart check box. This allows Solver to simultaneously run several GRG method solutions each starting at different and arbitrary chosen points. The downside of the GRG multistart computing approach is significantly more computational time. When choosing a multistart approach, the Population Size field indicates how many simultaneous GRG starts the Solver will create. If the value is left null or any number less than 10, then Solver starts 10 simultaneous runs. The starting point for each of these runs is defined from a number generator, which uses the value placed in the Random Seed field. When you place a value here, the multistart method uses the same starting points each time Solver starts. When you leave this field blank, Solver uses a different seed, which might yield a different (better or worse) value of the objective function every time you seek a solution.

Finally, the multistart approach requires you to choose whether to Require Bounds on Variables. If the check box is checked, the multistart method runs only if the lower and upper bounds are defined. Generally, the tighter the bounds on the variables, the better the multistart method is likely to perform.

Evolutionary Options

You can use the third tab in the Options window to select several options when an evolutionary solving method is used. Evolutionary methods, such as genetic algorithms, apply the principles of evolution found in nature to the problems of finding an optimal solution to a mathematical optimization model. In a nutshell, the evolutionary method consists of the following steps:

1. Create an initial set of possible solutions (initial population) and calculate the value of the objective function for each member of the initial population.

2. Select several members with good value of the objective function and apply mutation or crossover operators to generate a new generation of solutions.

3. Continue steps 1 and 2 until a satisfied solution is achieved.

When you use this approach, there are several options you can select: Convergence, Mutation Rate, Population Size, Random Seed, Maximum Time without Improvement, and Require Bounds on Variables. Again, you can use the default values for these options, unless you intentionally want to change the options, as explained in the following sections.

Convergence

This option is the same as the one in the GRG nonlinear method: The value entered in the Convergence field indicates the amount of relative change to be allowed in the last five iterations before Solver displays the Solver Converged to the Current Solution message and stops further attempts. Smaller values here normally mean that Solver takes more time, but stops at a point closer to the optimal solution.

Mutation Rate

The Mutation Rate option indicates the portion of the members of a given population that are altered or "mutated" to create a new trial solution, during each generation. In the Mutation Rate field, you must enter a number between 0 and 1. A higher mutation rate increases the diversity of the population and the chance that a new, better solution will be found; however, this might also increase total solution time.

Population Size

The Population Size option indicates the number of members in the population; that is, the number of different points or values for the decision variables. By default, the minimum population size is 10 members, which means that at any generation, the Solver will use a population size of 10 times the number of decision variables in the problem, but no more than 200.

Random Seed

The Random Seed option is used to generate a random choice in the evolutionary method. This field requires an integer value. When you enter a value in this field, the evolutionary method uses the same fixed seed; that is, the same choices each time Solver is run. If the field is left blank, the random number generator uses a different seed each time, which might yield a different final solution.

Maximum Time without Improvement

The Maximum Time without Improvement option indicates the maximum number of seconds that the evolutionary method continues without a meaningful improvement in the objective value. When you enter a value—let's say 30 seconds in this box—Solver will start to seek a better solution and will continue to improve the objective function. If the solution is not improved during any 30-second segment, Solver assumes that the optimal solution is reached and a message Solver cannot improve the current solution will display.

Require Bounds on Variables

The evolutionary method is more effective if upper and lower bounds on decision variables are defined. The tighter the bounds on the variables that you can specify, the better the evolutionary method performs. Selecting this check box constrains the evolutionary method to run only if bounds on all decision variables are defined.

Generate the Solution

After setting up constraints, objective function, and several Solver options, you simply click Solve (refer to Figure B-1) to generate a solution. As shown in Figure B-4, the Solver Results dialog box generates three reports. To select all three at once, either hold down the Ctrl key and click each one in turn or drag the mouse over all three. The three reports are generated in new sheets in the current workbook of Excel.

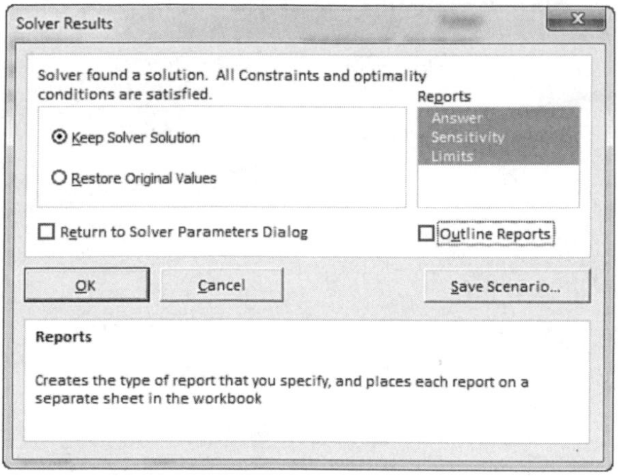

Figure B-4 Solver Results dialog box

References

[1] D. Laney, "3D Data Management: Controlling Data Volume, Velocity, and Variety," *Application Delivery Strategies*, 6 February 2001.

[2] Science Daily, "Big Data, for Better or Worse: 90% of World's Data Generated Over Last Two Years," 22 May 2013. [Online]. Available: http://www.sciencedaily.com/releases/2013/05/130522085217.htm. [Accessed 22 November 2013].

[3] The Economist, "Special Report," 27 February 2010. [Online]. Available: http://www.economist.com/node/15557443. [Accessed 22 November 2013].

[4] National Instruments, "The Moore's Law of Big Data," 15 January 2013. [Online]. [Accessed 22 November 2013].

[5] T. H. Davenport, "Analytics 3.0," *Harvard Business Review*, pp. 65–72, December 2013.

[6] D. Kiron, R. B. Ferguson, and P. K. Prentice, "From Value to Vision: Reimagining the Possible with Data Analytics," 5 March 2013. [Online]. Available: http://sloanreview.mit.edu/reports/analytics-innovation/. [Accessed 14 May 2014].

[7] J. Willwhite, "Getting Started in 'Big Data'," *Wall Street Journal*, p. B7, 4 February 2014.

[8] TMT Insight, "Big Data. What Is Your Strategy?," 2013. [Online]. Available: http://www.pinsentmasons.com/PDF/TMT-Insight-Big-Data-Autumn-2013.pdf. [Accessed 14 May 2014].

[9] D. Marks, "First Tennessee Bank: Analytics Drives Higher ROI from Marketing Programs," 9 March 2011. [Online]. Available: http://www.ibm.com/smarterplanet/us/en/leadership/firsttennbank/assets/pdf/IBMfirstTennBank.pdf. [Accessed 22 November 2013].

[10] C. Duhigg, "How Companies Learn Your Secrets," *New York Times*, 16 February 2012. [Online]. Available: http://www.nytimes.com/2012/02/19/magazine/shopping-habits.html?_r=0. [Accessed 22 November 2013].

[11] N. Golgowski, "How Target Knows When Its Shoppers Are Pregnant—and Figured Out a Teen Was Before Her Father Did," Mail Online, 18 February 2012. [Online]. Available: http://www.dailymail.co.uk/news/article-2102859/How-Target-knows-shoppers-pregnant--figured-teen-father-did.html. [Accessed 14 May 2014].

[12] T. H. Davenport and D. J. Patil, "Data Scientist: The Sexiest Job of the 21st Century," *Harvard Business Review*, October 2012. [Online]. Available: http://hbr.org/2012/10/data-scientist-the-sexiest-job-of-the-21st-century/ar/5. [Accessed 2014 May 2014].

[13] LinkedIn Press Center, "About LinkedIn," 2014. [Online]. Available: http://press.linkedin.com/about. [Accessed 14 May 2014].

[14] W. Winston, "Sports Analytics: Key Note Speech," in *Decision Sciences Institute's Annual Meeting*, Baltimore, MD, November 17, 2013.

[15] P. Andlinger, "RDBMS Dominate the Database Market, But NoSQL Systems Are Catching Up," 21 November 2013. [Online]. Available: http://db-engines.com/en/blog_post/23. [Accessed 5 July 2014].

[16] B. Betz, "Assessing Ticket Seller Fandango's Partnership with Yahoo Movies:Fandango's Reach Expands to Yahoo's 30 Million Users in the U.S.," 21 March 2012. [Online]. Available: http://investorplace.com/2012/03/assessing-ticket-seller-fandangos-partnership-with-yahoo-movies-cmcsa-yhoo/#.U8S9_fldVPM. [Accessed 15 July 2014].

[17] J. Jordan, "The Risks of Big Data for Companies," *Wall Street Journal*, 20 October 2013.

[18] McKinsey Global Institute, "Big Data: The Next Frontier for Innovation, Competition, and Productivity," June 2011. [Online]. Available: http://www.mckinsey.com/~/media/McKinsey/dotcom/Insights%20and%20pubs/MGI/Research/Technology%20and%20Innovation/Big%20Data/MGI_big_data_full_report.ashx. [Accessed 14 May 2014].

[19] D. Kiron, R. B. Ferguson, and P. K. Prentice, "Case Study: Oberweis Dairy," 5 March 2013. [Online]. Available: http://sloanreview.mit.edu/reports/analytics-innovation/case-study-oberweis-dairy. [Accessed 14 May 2014].

[20] L. Lawson, "Two Teens at 26?! What Big Data Got Wrong," 6 September 2013. [Online]. Available: http://www.itbusinessedge.com/blogs/integration/two-teens-at-26-what-big-data-got-wrong.html. [Accessed 14 May 2014].

[21] N. Silver, *The Signal and the Noise: Why So Many Predictions Fail—But Some Don't*, New York, New York: Penguin Group, 2012.

[22] E. Dumbill, "What Is Big Data?," O'Reilly Data, 11 January 2012. [Online]. Available: http://strata.oreilly.com/2012/01/what-is-big-data.html. [Accessed 14 May 2014].

[23] "107th Congress Public Law 204," U.S. Government Printing Office, 2002. [Online]. Available: http://www.gpo.gov/fdsys/pkg/PLAW-107publ204/html/PLAW-107publ204.htm. [Accessed 14 May 2014].

[24] M. Obrovac and C. Tedeschi, "Deployment and Evaluation of a Decentralised Runtime for Concurrent Rule-Based Programming Models," in *Distributed Computing and Networking: Lecture Notes in Computer Science*, vol. 7730, 2013, pp. 408–422.

[25] J. Dean and S. Ghemawat, "MapReduce: Simplified Data Processing on Large Clusters," December 2004. [Online]. Available: http://research.google.com/archive/mapreduce.html. [Accessed 5 July 2014].

[26] DSI Institute, "44th Annual Meeting: Decision Analytics— Rediscovering Our Roots," in *Proceedings of the DSI Annual Meeting*, Baltimore, MD, 2013.

[27] S. LaValle, E. Lesser, R. Shockley, M. S. Hopkins, and N. Kruschwitz, "Big Data, Analytics and the Path from Insights to Value," 8 September 2011. [Online]. Available: http://www.ana-lytics-magazine.org/special-reports/260-big-data-analytics-and-the-path-from-insights-to-value.pdf. [Accessed 19 May 2014].

[28] Z. MacDonald, "Teaching Linear Programming Using Micro-soft Excel Solver," *Computers in Higher Education Economics Review*, vol. 9, no. 3, 1995.

[29] G. B. Dantzig, *Linear Programming and Extensions*, Princeton, N.J.: Princeton University Press, 1963.

[30] J. Dongarra and F. Sullivan, "Guest Editor's Introduction: The Top Ten Algorithms," *Computing in Science and Engineering*, pp. 22–23, January/February 2000.

[31] G. Yarmish, H. Nagel, and R. Fireworker, "Recent Advances in Applications of Mathematical Programming to Business and Economic Problems," *Review of Business & Finance Studies*, vol. 5, no. 1, pp. 19–25, 2014.

[32] S. Sashihara, "From Big Data to Big Optimization: What Every Executive Needs to Know," *The European Financial Review*, 17 April 2012.

[33] S. Levy, "Jeff Bezos Owns the Web in More Ways Than You Think," 13 November 2011. [Online]. Available: http://www.wired.com/2011/11/ff_bezos/all/1. [Accessed 28 May 2014].

[34] E. Brynjolfsson, L. M. Hitt, and K. H. Heekyung, "Strength in Numbers: How Does Data-Driven Decision Making Affect Firm Performance?," April 2011. [Online]. [Accessed 28 May 2014].

[35] J. Cohen, B. Dolan, M. Dunlap, J. M. Hellerstein, and C. Welton, "MAD Skills: New Analysis Practices for Big Data," *Proceedings of the VLDB Endowment*, vol. 2, no. 2, pp. 1481–1492, 2009.

[36] M. Avriel, *Nonlinear Programming: Analysis and Methods*, Mineola, New York: Courier Dover Publications, 2003.

[37] M. S. Bazaraa and C. M. Shetty, *Nonlinear Programming: Theory and Algorithms*, New York: John Wiley & Sons, 2006.

[38] G. P. McCormick, *Nonlinear Programming: Theory, Algorithms and Applications*, New York: John Wiley & Sons, 1983.

[39] I. Griva, S. G. Nash, and A. Sofer, *Linear and Nonlinear Optimization*, Philadelphia: Society for Industrial and Applied Mathematics, 2008.

[40] D. G. Luenberger and Y. Ye, *Linear and Nonlinear Programming*, 3rd ed., New York: Springer, 2008.

[41] F. W. Harris, "How Many Parts to Make at Once," *Operations Research*, vol. 38, no. 6, pp. 947–950, 1990 [Reprint from 1913].

[42] FICO, "FICO Helps Enterprises Tackle Big Data Optimization Problems," 26 January 2014. [Online]. Available: http://www.fico.com/en/about-us/newsroom/news-releases/fico-helps-enterprises-tackle-big-data-optimization-problems/. [Accessed 1 June 2014].

[43] A. Charnes, W. W. Cooper, and R. Ferguson, "Optimal Estimation of Executive Compensation by Linear Programming," *Management Science*, vol. 1, pp. 138–151, 1955.

[44] S. M. Lee, *Goal Programming for Decision Analysis*, Philadelphia: Auerback, 1972.

[45] J. P. Ignizio, *Goal Programming and Extensions*, Lexington: Lexington Books, 1976.

[46] C. Romero, *Handbook of Critical Issues in Goal Programming*, Oxford: Pergamon Press, 1991.

[47] M. S. Schniederjans, *Goal Programming Methodology and Applications*, Boston: Kluwer Publishers, 1995.

[48] D. F. Jones and M. Tamiz, "Goal Programming in the Period 1990–2000," M. Ehrgott and X. Gandibleux, Eds., Kluwer, 2002, pp. 129–170.

[49] D. F. Jones and M. Tamiz, *Practical Goal Programming*, New York: Springer Books, 2010.

[50] T. H. Davenport and J. Dyché, "Big Data in Big Companies," May 2013. [Online]. Available: http://www.sas.com/resources/asset/Big-Data-in-Big-Companies.pdf. [Accessed 3 June 2014].

[51] M. Nir, J. Bloomberg, T. Arthur and P. Roma, "Optimization Toolbox Adds Mixed-Integer Linear Programming for MAT-LAB," 5 June 2014. [Online]. Available: http://dotnet.sys-con.com/node/3100697. [Accessed 5 June 2014].

[52] FICO, "FICO® Xpress Optimization Suite," 2014. [Online]. Available: http://www.fico.com/en/products/fico-xpress-optimization-suite/. [Accessed 5 June 2014].

[53] AIMMS, "IBM ILOG CPLEX Solver," 2014. [Online]. Available: http://www.aimms.com/aimms/solvers/cplex. [Accessed 5 June 2014].

[54] Gurobi Optimization, "Gurobi 5.6 Performance Benchmarks," 2014. [Online]. Available: http://www.gurobi.com/pdf/Benchmarks.pdf. [Accessed 5 June 2014].

[55] R. Handfield, F. Straube, H.-C. Pfohl, and A. Wieland, "Trends and Strategies in Logistics and Supply Chain Management," 19 June 2013. [Online]. Available: http://scmresearch.org/2013/06/19/trends-and-strategies-in-logistics-and-supply-chain-management/. [Accessed 7 June 2014].

[56] M. Jesk, M. Grüner and F. Weiß, "Big Data in Logistics: A DHL Perspective on How to Move Beyond the Hype," December 2013. [Online]. Available: http://www.dhl.com/content/dam/downloads/g0/about_us/innovation/CSI_Studie_BIG_DATA.pdf. [Accessed 7 June 2014].

[57] J. Mawhinney, "Marketing Analytics vs. Business Analytics: What's the Difference (And Who Should Care)?," 6 February 2014. [Online]. Available: http://blog.hubspot.com/insiders/marketing-analytics-vs-business-analytics. [Accessed 9 June 2014].

[58] J. DeMers, "2014 Is the Year of Digital Marketing Analytics: What It Means for Your Company," 10 February 2014. [Online]. Available: http://www.forbes.com/sites/jaysondemers/2014/02/10/2014-is-the-year-of-digital-marketing-analytics-what-it-means-for-your-company/. [Accessed 9 June 2014].

[59] P. S. Fader, B. G. S. Hardie, and K. L. Lee, "RFM and CLV: Using Iso-Value Curves for Customer Base Analysis," *Journal of Marketing Research*, vol. 42, no. November, pp. 415–430, 2005.

[60] R. Venkatesan and V. Kumar, "A Customer Lifetime Value Framework for Customer Selection and Resource Allocation Strategy," *Journal of Marketing*, vol. 68, no. October, pp. 106–125, 2004.

[61] W. J. Reinartz and V. Kumar, "On the Profitability of Long-Life Customers in a Noncontractual Setting: An Empirical Investigation and Implications for Marketing," *Journal of Marketing*, vol. 64, no. October, pp. 17–35, 2000.

[62] J. Vein, "An Advanced Marketing Analytics Cheat Sheet for CEOs," 19 May 2014. [Online]. Available: http://www.forbes.com/sites/forbesinsights/2014/05/19/an-advanced-marketing-analytics-cheat-sheet-for-ceos/. [Accessed 10 June 2014].

[63] T. Davenport, "Big Data and the Role of Intuition," 24 December 2013. [Online]. Available: http://blogs.hbr.org/2013/12/big-data-and-the-role-of-intuition/. [Accessed 10 June 2014].

[64] F. Germann, G. L. Lilien, and A. Rangaswamy, "Performance Implications of Deploying Marketing Analytics," *International Journal of Research in Marketing*, vol. 30, no. 2, pp. 114–128, June 2013.

[65] W. L. Winston, *Marketing Analytics: Data-Driven Techniques with Microsoft Excel*, Indianapolis, IN: John Wiley & Sons, 2014.

[66] IBM Corporation, "First Tennessee Bank: Analytics Drives Higher ROI from Marketing Programs," 2011. [Online]. Available: http://www.ibm.com/smarterplanet/us/en/leadership/first-tenbank/assets/pdf/IBM-firstTennBank.pdf. [Accessed 11 June 2014].

[67] T. D'Auria, "The Fall of RFM Analysis: Why Recency, Frequency and Monetary Value Are Just Not Getting It Done Anymore for Marketers," 19 June 2009. [Online]. Available: http://www.imn-unlocked.com/bi-creativecomputing/e_article001467796.cfm?x=b11,0,w. [Accessed 12 June 2014].

[68] C. R. Harrell, R. E. Bateman, T. J. Gogg, and J. R. A. Mott, *System Improvement Using Simulation*, 3rd ed., Orem, UT: PROMODEL® Corporation, 1995.

[69] T. O'neal, "Simulation in the Big Data Era," 3 June 2014. [Online]. Available: http://www.supercomputingonline.com/latest/topics/this-month/58004-simulation-in-the-big-data-era. [Accessed 12 July 2014].

Index